# technology and the environment in history

## Technology in Motion

Pamela O. Long and Asif Siddiqi, Series Editors

Published in cooperation with the Society for the History of Technology (SHOT), the Technology in Motion series highlights the latest scholarship on all aspects of the mutually constitutive relationship between technology and society. Books focus on discrete thematic or geographic areas, covering all periods of history from antiquity to the present around the globe. These books synthesize recent scholarship on urgent topics in the history of technology with a sensitivity to challenging perspectives and cutting-edge analytical approaches. In combining historical and historiographical approaches, the books serve both as scholarly works and as ideal entry points for teaching at multiple levels.

# technology
# and the
# environment
# in history

*Sara B. Pritchard*
*Carl A. Zimring*

 Johns Hopkins University Press
Baltimore

Johns Hopkins University Press
2715 North Charles Street
Baltimore, Maryland 21218–4363
www.press.jhu.edu

Library of Congress Cataloging-in-Publication Data

Names: Pritchard, Sara B., 1972– author. | Zimring, Carl A., 1969– author.
Title: Technology and the environment in history / Sara B. Pritchard, Carl A.
    Zimring.
Description: Baltimore : Johns Hopkins University Press, [2020] | Series:
    Technology in Motion | Includes bibliographical references and index.
Identifiers: LCCN 2020002952 | ISBN 9781421438993 (paperback ; acid-free
    paper) | ISBN 9781421439006 (ebook)
Subjects: LCSH: Industries—Environmental aspects—History. | Environmental
    technology—History. | Human ecology. | Pollution—Health aspects.
Classification: LCC TD194 .P75 2020 | DDC 304.209—dc23
LC record available at https://lccn.loc.gov/2020002952

A catalog record for this book is available from the British Library.

*Special discounts are available for bulk purchases of this book. For more
information, please contact Special Sales at specialsales@press.jhu.edu.*

Johns Hopkins University Press uses environmentally friendly book materials,
including recycled text paper that is composed of at least 30 percent postconsumer waste, whenever possible.

# Contents

# Acknowledgments

We thank Pamela O. Long and Asif Siddiqi, current coeditors of this series, for their interest in this project, long-standing support, and patience as we brought this book to fruition. We are also grateful to Matt McAdam and Johns Hopkins University Press for their support of the series in general and our book in particular.

We acknowledge Hugh Gorman and Dolly Jørgensen, each of whom read an early version of the chapters and provided essential, insightful feedback. Robert Kulik was a great editor, as always.

We are indebted to the scholars working in the history of technology, environmental history, and their intersection—a subfield now known as *envirotech*. Our brief syntheses here do not do full justice to their scholarship and contributions. Please see the notes and bibliography for just a selection of their work.

We note that this book is being published in 2020—the twentieth anniversary of the founding of Envirotech, a special interest group (SIG) affiliated with the Society for the History of Technology and the American Society for Environmental History; it also meets at the European Society for Environmental History. The Envirotech SIG has been a crucial catalyst for scholarship at the nexus of these fields, fostering scholarly conversations and networks, providing a professional home, and offering a welcoming community.

# technology and the environment in history

# Technology and the Environment in History

S cientists, policymakers, politicians, and citizens are confronted by dilemmas at the nexus of technology and the
environment. Consider carcinogens, endocrine disrupters, and
"superbugs." Or energy crises—from fossil fuels to nuclear
power—and failing infrastructure of modernity, such as water
and sewer systems and transportation networks. In the 2010s,
debates such as those centering on the privatization of water in
Latin America, lead contamination in Flint, Michigan (USA),
and construction of the Keystone XL Pipeline to send oil from
Alberta (Canada) to refineries and distribution centers in the
United States sparked environmental activism and social movements. Still other concerns frequently make it into bold head-
lines: global climate change, genetically modified organisms, and
the Anthropocene. Of course, these debates have deep roots and
longer histories.

Historians are also increasingly interested in technology
and the environment. Environmental historians and historians
of technology have taken the lead in studying these topics. They

have examined case studies and developed conceptual tools to help us think about technology and the environment in the past as well as the present. Their work complements, but sometimes challenges, established popular and historical narratives. For instance, both fields have asked big questions about agency: in other words, who—or even *what*—drives historical change? Environmental history and the history of technology have also influenced such related disciplines as environmental anthropology, geography, political ecology, and science and technology studies (STS).

This book offers an overview of scholarship on technology and the environment in history.[1] We seek to synthesize, highlight, and discuss some of the important topics, historical processes, and scholarly concerns that have emerged from recent work in thinking about both technology and nature in the past. Our discussion builds on the research of a number of scholars and literatures. We are particularly indebted to those working at the intersection of environmental history and the history of technology—a nexus now called *envirotech*. At the same time, we look outward and forward in this book. We identify several promising areas in more-formative stages of research and call attention to conversations in related fields, such as discard studies, posthumanism, and sensory studies, which have flourished over the past decade.

We adopt an *analytic* approach to history in this book. There are many different approaches to studying and writing history. Historians have often used geography or chronology to organize their understanding of the past—studying the history of modern Japan or pre-1865 US history, for example. Analytic approaches such as economic history, gender history, the history of sexuality, and—especially relevant to this book—environmental history and the history of technology offer different perspectives on the past. Rather than organizing a study by place (often a nation-state) or time period (such as medieval or modern history), an analytic perspective allows us to examine important historical phenomena that do not necessarily follow political borders or the convenient markers of centuries. It thus facilitates an examination of historical events and processes across time and place. Moreover, an analytic approach provides a particular lens or perspective on the past, bringing a specific thematic concern (economics, gender, technology, and so on) into sharper relief. For the purposes of this book, an analytic approach to history helps us identify and analyze how technology and the environment, as well as their complex dynamics, both reflect and explain significant historical change.

There are several challenges to writing this kind of history. First, our goal is *not* to be comprehensive. That is an impossible task. This book is partial in both senses of the word, and it undoubtedly reflects our own research interests and expertise. For instance, North America, Europe, the modern era, and capitalist economies are overrepresented here. We certainly hope that future scholars will not only build on but also revise our work (as well as the scholarship we draw upon) to make historical analysis of technology and the environment more global, culturally inclusive, and comparative in scope.[2]

Doing so is important for several reasons. For starters, the very categories of *technology* and *environment* do not necessarily make sense in all cultural, historical, and linguistic contexts.[3]

More significant, global-scale processes such as empire, capitalism, and industrialization—with environmental and technical factors vital to all three—unevenly affected the global North and South. Until recently, the history of technology tended to focus on more-traditional historical actors, sites, and processes. Many studies were defined around large-scale technological systems, which usually require extensive capital, expertise, and centralized power to create. Other studies, influenced especially by social and gender history, have defined technology more broadly—as the knowledge, skills, and tools of making or doing things—while still others have emphasized the persistence of "older" technologies.[4]

While we often think of technology in terms of innovation and novelty, maintenance, repair, and established technologies may actually be more representative of our daily lives and experiences.[5] These insights have crucial implications for envirotech scholarship, especially in and about the global South. They draw attention to technologies of environmental management that may be relevant to an enormous proportion of the world's population; continuity (for many) frequently persists amid seeming dramatic change (for a minority). Even "high" technology, like atomic bombs, simultaneously depended on technologies and technological actors (uranium ore, African miners, hand tools) not usually considered as such.[6]

A critical question, then, is how envirotech analyses to date—including the historical processes, chronology, and concepts identified and developed thus far, which have been based largely on sites in North America and Europe—illuminate (or actually fail to explain) those in the global South. At the same time, we do not want to fall prey to a tidy divide between global North and South, or necessarily assume complete differences between the developed and developing world. For example, examining water systems in diverse historical and cultural settings undoubtedly suggests a common dependency on weather and climate regardless of the time, place, or technology. Nonetheless, vulnerabilities to, say, drought differ depending on a group's ability to buy bottled water or pay utilities to repair problems more quickly. In other cases, the urban and rural may better explain differences than "global North" and "global South." For instance, networks that supply Johannesburg (South Africa) and Los Angeles (USA) with water may have significant similarities, while water collection methods in rural China may have little in common with those in Beijing. Again, we hope future scholars will consider technology and the environment in wider cultural, historical, and political contexts, and develop and refine analytic tools to better understand more of the human experience in all of its complexity and diversity.

We begin with a brief discussion of some of the major literatures, historiographical trends, and concepts that scholars have developed for thinking about technology and the environment in history. The six chapters that follow are organized thematically. This organization is admittedly artificial. As the chapters show, there are many connections among them, both historical and theoretical. Distinct chapters risk obscuring these links, but they are a necessary and practical convenience.

Chapter 1 considers *food and food systems*, which are obviously fundamental to humanity. By focusing on technology and the environment, we begin to understand how humans have manipulated organisms and ecosystems to produce nutrients for societies throughout history, albeit in diverse ways.

Many modern food systems are highly industrial, leading us to the next chapter's theme: *industrialization*. Historical scholarship on industrialization has often focused on technological and social change. Newer work has emphasized not only industrialization's ecological impacts but also the ways in which environmental processes constrained industrialization and required shifts in the relationships between human and nonhuman nature.

Historians of industrialization have frequently focused on production and consumption, but some scholars now emphasize the afterlife of industrial things. This is the focal point of discard studies—an interdisciplinary field that considers the creation, classification, management, disposal, and diverse forms of waste, as well as its consequences for ecosystems, humans, and other species. Chapter 3 examines *discards* in their multifaceted forms, complex histories, and also sometimes-unexpected possibilities.

The fourth chapter turns to *disasters*, which are complex, global, and normal in industrial economies and societies. Disasters are especially effective at exposing the fallacy of tidy divisions among nature, technology, and society.

We then explore *the body* in chapter 5, considering how environmental and technological change shape cultural understandings of the human body and cause it both intended and unintended physical transformations. Bodies also reveal the porous boundaries among technology, the environment, and the human.

The final chapter analyzes *sensescapes* and how environmental and technological change have reshaped humans' (and potentially nonhumans') sensory experiences over time.

## Concepts

Over the past several decades, historians and other scholars have developed new concepts and theories to understand, describe, and explain interactions between technology and the environment in the past. Such concerns are not entirely new. For example, environmental anthropologists have studied how cultures relate to nonhuman nature. In the 1960s, anthropologist Marvin Harris specifically theorized the "techno-environment"—the application of technology to the environment.[7] More recently, environmental historians, historians of technology, STS scholars, and others have investigated diverse case studies with the aim of making broader generalizations about how people, the environment, and technology interact with one another. Review essays in these fields provide more-detailed discussions of these issues.[8]

Here, we focus on five key arguments and concepts that are relevant to the chapters that follow: (1) porosity, or the permeable boundaries among technology, the environment, and society; (2) systems—specifically, envirotechnical systems; (3) hybridity; (4) biopolitics; and (5) environmental (in)justice. Highlighting these concepts is a second way we pursue an analytic approach to technology and the environment in the past.

For centuries, the Western tradition has generally neatly separated nature and technology.[9] Since at least the eighteenth century, and particularly since World War II, the actual world has little resembled this ideal.[10] The definitions and borders of the environment, technology, and society are ever murkier. Historians and scholars in related fields have developed new ways of thinking in response to this increasingly complex, messy world.

For instance, scientists believe climate change is altering the ratio of male and female sea turtles born in Australia's Great Barrier Reef, thereby fundamentally challenging the seemingly fixed biological category of sex.[11] Extensive flooding in Chennai (India) and New York City (USA) literally erode the boundaries of the environment and technological infrastructure. In addition, many case studies demonstrate that the dynamics among technological, social, and environmental change are more complex than simply a single factor causing the others to change. This understanding of causality is sometimes called technological determinism, or environmental determinism. We therefore need a more nuanced, sophisticated toolkit to understand these dynamic processes in their complicated, interactive detail.[12] New conceptual tools seek to better represent these relationships while fostering new responsibilities and ethics. In short, an increasingly complex socio-enviro-technological world requires new ideas, new ways of thinking, and therefore new scholarly tools.

This initial insight—that the boundaries of social, technological, and environmental things and processes are often unclear and porous—is a critical contribution that environmental historians and historians of technology offer scholars and citizens alike.[13] These scholars also underscore the fuzzy boundaries between the (human) body and what scientists call the ecosystem. As Kroll-Smith and Lancaster write, we typically think of the human body in individual, bounded terms—"my body." In contrast, "the environment" is something "out there," external to the human body and distinct from it. Moreover, as these scholars note, "the pronoun 'my' in front of 'body' signals a possessive interest in human bodies that differs from the typical article 'the' that precedes 'environment.'"[14]

Yet, as recent environmental historians and historians of medicine have shown, lead, mercury, radiation, and endocrine disruptors, among other substances, can have profound effects on both human and nonhuman bodies.[15] What we put into our bodies—antibiotics, birth control pills, and so on—can end up in the wider environment, which can, in turn, affect us. And what we put into nonhuman bodies, such as poultry and livestock raised for meat production, can also affect humans. Recent scientific studies of the human microbiome suggest that the human body is an ecosystem *within* an ecosystem.[16] Neither system, however, is closed. One can affect the other.

Indeed, since the 1980s, historians of technology have used the concept of *systems*, initially formulated by Thomas Parke Hughes, as one way to describe these kinds of interconnections, relationships, and feedback loops. Hughes focused on large-scale "technological" systems, but argued that they were shaped by social, political, and economic factors, not just technical ones. In his view, the environment was external to those systems. Later scholars built on and revised Hughes's ideas, arguing that the environment is also part of "sociotechnical" systems.[17]

The related concept of *envirotechnical systems* instead emphasizes how environmental and technological systems both shape and are shaped by each other.[18] For example, hydroelectric dams may transform rivers, at times radically, but they are still ultimately dependent on the rivers' flow. Low and irregular flows minimize energy production or the distribution of water through irrigation networks or municipal water supply systems. Moreover, despite extensive efforts to dam, dredge, and dike rivers, they still flood. Rivers may be rationalized and managed. They are often subject to large-scale intervention by people, through technology, to serve human ends. In this sense, rivers are technological. But they are still natural systems subject to weather, climate, sedimentation, and a multitude of other environmental processes. The concept of *envirotechnical* system seeks to capture the ways in which objects, artifacts, and systems are both natural and technological.

As the term *envirotechnical system* itself suggests, environmental historians have been influenced by the idea of *hybridity* as a way to depict phenomena that are at once social, ecological, and technological.[19] Some historians have referred to hybrid landscapes, and others have argued that hybrid causality better explains historical change.[20] These analytic tools challenge the clear separation of technology and the environment, both as cultural ideas and material things. Instead, they stress interconnections and entanglements, multiplicity and simultaneity. The concepts push scholars and citizens to consider interdependency, unintended consequences, and shared fates, rather than strict divisions between things often held apart.

*Biopolitics*, another important concept premised on the idea of hybridity, is also relevant to the following chapters. The French theorist Michel Foucault originally proposed the concept of biopolitics, which explores how the state analyzes, manages, and intervenes in life. Foucault was particularly interested in governmental surveillance, studies, and regulatory regimes analyzing human populations—from prisons and medical patients to citizens counted in the census.[21] We can also adopt a broader definition of *bio-* to consider nonhuman forms of life, and the ways in which the state, bureaucracies, and experts manage how people use the environment—and thereby manage the environment itself. Thus, it is not just humans who are subject to biopolitics, but also nonhuman beings, as well as nonliving entities.[22]

Foucault's concern with power anticipates a final major theme: *environmental justice*, and its opposite. Since the early 1980s, activists, social movements, and activist-scholars have studied environmental racism: the disproportionate environmental burdens the poor, communities of color, and other marginalized groups bear. Reshaping diverse environments through technological artifacts and systems has often resulted in unequal effects on social groups within and across societies. Scholars in technology studies have argued that we can think about technology as politics: the material enactment and assertion of power. In this sense, it is vital to consider how technological change can reproduce, reinforce, and even amplify environmental injustice at multiple scales.

## Chronology

As mentioned, we have adopted an analytic approach in this book, organizing the following chapters around significant themes and highlighting a number of concepts that can help us think about specific examples in a broader way. Nonetheless, we pause here to make a few comments about chronology, noting that some chapters have a general chronological sweep; in others, individual sections are organized roughly chronologically.

To date, envirotech scholars have not developed or proposed a chronology to describe technology and the environment in the past. Here, we tentatively suggest some key processes, moments, and periodization.

If we take the long view and consider the whole of human history, the ability to control fire is a crucial moment in a global history of technology and the environment. Fire is a technology. Planned burns had important ecological effects and provided humans with the ability to cook calorie- and nutrient-rich food, which in turn had significant implications for the evolution of humans. Much more recently, the domestication of plant and animal species—turning nature into technologies—also served human ends.

Still more recently, urbanization, even when not on the scale of modern megalopolises, required hinterlands to support concentrated human populations. In this case, various technologies provided direct and indirect support. Transportation networks, transit technologies, and various systems for supplying water, food, and fuel and for disposing of waste enabled more people to live in closer quarters.

Colonialism, slavery, capitalism, and industrialization had transformative historical consequences in many domains between the fifteenth and nineteenth centuries. These intertwined processes facilitated, expanded, and accelerated contact and exchanges of various sorts across large parts of the globe. For one, they widened and intensified resource extraction. Colonial subjects and enslaved peoples were treated much like natural resources or inputs. A key issue is not just that networks of environments and technologies were linked in time and space in unprecedented ways through new forms of political economy, but also *how* and *why* they were so linked. In other words, who benefited from the forging of envirotechnical systems central to the modern world?

Even as we study how these processes, collectively and cumulatively, reflected and resulted in significant changes in human societies, the environment, and technology, we also need to pay attention to absences, continuities, and exceptions. For instance, many narratives of electrification focus on significant changes in the late nineteenth and early twentieth centuries. Yet even in "electrified" societies, many parts of rural Europe and North America did not have electricity until the 1960s. Meanwhile, parts of the developing world are now leapfrogging the installation of electrical systems based on nineteenth- and twentieth-century technologies, to instead develop an electrical industry based on solar power. In another example, metropolitan infrastructure for water, sewers, or energy is all too often concentrated in wealthy centers, while surrounding, poorer neighborhoods or "slums" lack access to it. At the same time, some of these communities demonstrate considerable creativity and innovation by tapping available networks, even if their access is temporary, limited, unpredictable, and ultimately subject to sanction by utilities or the government.

A key feature of industrialization was growing reliance on

fossil fuels, another key moment in this chronology. Coal and, later, oil and gas were fundamental to the industrial age. Using fossil fuels tapped energy sources at an unprecedented rate. Numerous technological artifacts and systems based on fossil fuels could do more work faster than with energy derived from animals, people, water, or wind. The speed and scale of production had implications certainly for industrial workers, but also for the resources extracted in colonies and eventually for the climate at a global scale.

Synthetic chemistry and the atomic age have roots in the nineteenth century, but they really developed and expanded during and after World War II. Both generated new materials, new kinds of pollution, and vast geographies of exposure well beyond corporate or national borders. The movement of air, soil, and water made them global problems.

One possible end to this periodization is the coining of the term *Anthropocene* in 2000 to emphasize the role of humans in reshaping planetary processes to a historically unprecedented extent. Although scholars from a wide range of fields have debated the term and offered alternatives, many feel the term is justified as a definition of a new geologic era. Yet, as many critics have noted, the Anthropocene is based on a universal, undifferentiated "(hu)man" (*anthro-*) and therefore fails to acknowledge the ways in which resource use and, ultimately, responsibility have been—and remain—unevenly distributed across humanity.[23] Regardless of the debate over terms, and of dating when the Anthropocene began, this concept hints at the temporal and geographical scale of "anthropogenic" environmental change. It thus raises crucial questions about technology and the environment not only in the past but, more significant, in the present and future. In this sense, thinking with the Anthropocene offers valuable opportunities to reflect upon better, more-just ways whereby people, technology, and the environment can coexist.

Historians and scholars working in related disciplines have developed the foregoing and other concepts. However, they are not abstract academic exercises. These ideas help us understand, describe, and explain how technology and the environment have interacted not only in the past but also in the present. They also have design and policy implications. For example, the ongoing water crisis in Flint, Michigan, is in part a story of old lead pipes and a contaminated river diverted through aging infrastructure. Lead contamination is less surprising, though, if we work from the concept of envirotechnical systems. The conjuncture of nature and technology in this case has significant consequences for the communities exposed to lead, especially their children, and foremost poorer children of color. The differential effects of lead on "people" in Flint thus illustrates the concept of environmental injustice.

By thinking about technology and the environment at the same time, we challenge clean divisions between environmental and technological processes, and instead assume that technical systems are never entirely closed systems. Doing so enables us to expect, anticipate, and hopefully prevent some of their negative consequences for both people and nonhumans. Envirotechnical thinking therefore encourages us to assume inextricable connections among social, ecological, and technological processes. In this way, envirotechnical perspectives might help us design, engineer, manage, and otherwise engage with the surrounding world in ways that are, we hope, more sustainable and just for both humanity and the planet.

# Food and Food Systems

E ating is a natural activity. Like breathing air and drinking water, eating food is vital to sustaining life. These ostensibly natural acts, however, have envirotechnical dimensions both past and present. The act of breathing is made possible in unforgiving environments (high altitudes, bodies of water, space) by the storage and delivery of pressurized oxygen. Drinking water is rendered safe and abundant by several technical systems, including sewage treatment, chemical disinfection, and desalination. The food we eat is subject to a complex and varied set of relationships between technologies and the environment.[1]

We may see that food is subject to these relationships when we go to a store and purchase groceries that traveled hundreds or even thousands of miles from their point of origin. We also know that highly processed foods such as candy may have colors, textures, and tastes that do not appear in nature. These are just two present-day examples of the complex interactions between nature and technology in food in the industrialized world, and they reflect a long history of such interactions that scholars have investigated in recent years.

The history of food is in many respects the history of human experience. Many of the major changes in human history—the development of nomadic tribes, the rise of agricultural societies, the emergence of civilizations, the increasingly broad scope of trade and exchange, and the Industrial Revolution—are results of human endeavor to increase food security.

Much of this history may be understood as the history of food *systems*, the process of humans negotiating the opportunities that nature and technology afford for sustained human benefit. Food systems comprise the processes and infrastructure involved in bringing food to a society. These steps may include the growing, harvesting, processing, preserving, packaging, transporting, consumption, and disposal of food. An envirotechnical approach to the history of food systems allows us to understand the myriad ways in which we humans have fed ourselves and to assess how we have cultivated and manipulated organisms, the materials harvested from organisms, and the land and water used to develop crops and livestock. The systemic organization of these manipulations has provided abundant, attractive, and delectable foods for the world's peoples since the advent of river valley civilizations.

## *Technology and the Rise of Agriculture*

Eating has involved mediations between technology and nature since the earliest societies, and changes in these mediations have had significant effects on human and environmental history. Until approximately 10,000 BCE, the vast majority of human societies were organized as hunters and gatherers. These societies were nomadic and small (perhaps two dozen humans), and all members were involved in producing food for their group. Because these societies traveled to where food sources existed, hunter-gatherers built few permanent structures.[2]

Hunter-gatherer societies still exist, but were succeeded by agricultural societies as the dominant form of human organization during the Neolithic Revolution. The transition to the domestication of plants and animals began in river valleys in Mesopotamia (now the Middle East, also known as the Fertile Crescent) and elsewhere. Mesopotamian farmers domesticated barley and wheat, then lentils, peas, chickpeas, and flax.[3]

Agriculture still required most members of a society to produce food; however, the conditions and scale of food production changed. Agricultural societies, unlike those of hunter-gatherers, required tending a fixed plot of land over a succession of seasons. This food system depended on fertile soil, adequate water, and human and animal labor to function; it was vulnerable to drought, pestilence, and war. Yet it produced far more-stable systems of supplying food to the population, and the population of agricultural societies grew much larger. Fixed communities in river valleys gave rise to civilizations in the Middle East, India, and China—and with them, new technologies, new agricultural practices, and new wars, as fixed societies defended their crops and livestock from invaders. The transition from hunting and gathering to agriculture was gradual, with agriculture emerging as the dominant form of global food production by 2500 BCE. Using an environmental determinist approach, Jared Diamond argues that East Asia and Europe developed agriculture ahead of sub-Saharan Africa and Central Asia because of their location in temperate climates (with suitable plants and animals) and in relatively remote terrain, safe from invasion.[4]

Agriculture in the Americas involved the domestication of crops distinctly different from those in Africa, Asia, or Europe, and the food systems of the Americas did not interact with the rest of the world until 1492. Domesticated crops in the Americas included maize, potatoes, and squashes. The cultivation of agriculture by indigenous peoples involved several manipulations of the land, including the selective application of fire in forests to till soil, and uses of polyculture (agriculture in which more than one species is grown) that English colonists supplanted in replicating their homeland's "world of fields and fences" in order to intensively produce particular crops.[5]

The rise of agriculture reshaped societies and provided a context for new technological innovations as humans cultivated and manipulated species of plants and animals. Although many of these are beyond the scope of this chapter, major technological advances in food production included the creation of water wheels to irrigate fields in Mesopotamia and, over time, the advent of iron sickles and ploughs to till soil. These tools were often used in conjunction with animals to power the planting and harvesting of agricultural crops. Indeed, the use of animal labor is an ongoing theme in envirotechnical history. For example, humans altered horses' physiology to maximize their use in transportation and industry, transforming the animals into "living machines."[6]

## The Columbian Exchange

Technological innovation in food systems accelerated with advanced transportation technologies. The successful (if initially inadvertent) linking of Eurasia and the Americas by wooden sailboats linked the agricultural crops of the Americas with the existing food systems in other parts of the world. The establishment of transatlantic trade had vast consequences for food systems, bringing New World crops not only to Europe but to Africa and Asia as well.

The influx of maize, squash, and other New World foods reshaped agricultural production in Africa. Between the sixteenth century and the end of the twentieth century, maize accounted for more than half the calories ingested by people in several African nations, as the crop supplanted the continent's indigenous grains of sorghum, millet, and rice. Such transformations altered the makeup of human bodies; they also contributed to the reduction of biodiversity in ecosystems across Earth.[7]

As crops from the Americas transformed food systems in Europe, European agricultural practices transformed land use in the Americas. English farming practices reshaped the land in colonial Massachusetts. A significant technological innovation was the use of fences to restrict one farmer's land from another's, thereby establishing private property that could be used for growing commodity crops. These human structures made of wood interrupted landscapes and posed barriers to migrating animals, producing hierarchies on the land that placed human needs above those of other species. (The monocultural production this organization would generate unintentionally prioritized the needs and wants of animals that could breach fences and thrive on the vast acreages of food. Humans subsequently attempted further technological innovations to retain dominance over the land; some of these attempts are explored later in this chapter.)[8]

European emphasis on commodity crops shaped the growth of plantations so as to maximize the value of New World crops, shaping food systems in ways that were crucial to the development of industrial systems. Sugar production in the Caribbean, Central America, and South America may be seen as an "agro-industry," blending industrial methods and organization of labor in an agricultural environment. This food system not only manipulated the land to produce increased yields of a commodity crop, but also transformed human labor forcibly into "cogs in the machine" that produced a commodity for consumption an ocean away from where it was grown.[9]

The food thus produced transformed bodies across Europe by providing affordable, calorie-dense nutrients for the growing class of industrial, urbanized wage laborers in cities from London (UK) to Moscow (Russia). Wooden ships facilitated the global trade in sugar, spices, and salt, additives that could alter the caloric density, durability, and taste of foods.[10]

## Industrialized Food Systems

The global sugar trade prefigured major changes in food systems rendered by the Industrial Revolution. Since the early nineteenth century, the use of fossil fuels and mechanized har-

This combined barley harvester and thresher from 1917–1918, shown "in action on a big ranch in southern California," exemplifies an industrial food system in the United States. *US National Archives and Records Administration (31482104)*

vesting, processing, and distribution technologies have transformed the foods we eat.

Although envirotechnical relationships have formed food systems throughout human history, the complexity and scale of the negotiations between technology and nature expanded with the advent of industrialized food systems in the nineteenth and twentieth centuries. Certainly much of the attention that scholars of technology and nature have given to food systems focuses on this period, and for good reason. Mechanical, chemical, and genetic advances have transformed modern food systems into technological artifacts distinct from their preindustrial precedents. Those same advances have produced environmental consequences that are distinct in both degree and kind.

William Cronon's environmental histories show the transitions of food systems in the Americas. His *Changes in the Land* depicts the transition from polyculture food production to more intensive monoculture in New England during the colonial era. The rise of industry revolutionized food production. For the first time, the majority of a society's people need not focus on producing food for that society. The successful harnessing of fossil fuels and machinery to cultivate the land, harvest crops, slaughter animals, and preserve and transport food reshaped food systems and societies.

Cronon's *Nature's Metropolis* depicts the advent of an industrialized, national food system that transported animals and grains from across the midwestern United States to Chicago via rail. There, the animals were slaughtered at a rate approaching 70,000 per day and turned into meat. Meat and grains then left Chicago via rail to produce a standardized set of foods available across the United States.[11]

This broad story has many particulars that involve the dynamic intersection of nature and technology. A national food system involving meat required the innovation of refrigeration for homes and rail cars. In the northern Great Lakes region, an industry emerged focused on harvesting ice. Grain requirements spurred innovation in storage and transportation, including standardized rail cars and grain elevators. The traffic of commodity foods by rail required adherence to precise clock time to

Chicago was home to the advent, in the 1860s, of an industrialized national food system that transported animals and grains from across the midwestern United States to the city via rail. The Union Stock Yards on Chicago's South Side was the final destination of animals transported to Chicago. There, the animals were slaughtered and turned into meat in mechanized slaughterhouses. The same railroads then distributed meat and grains to cities across the country. *Library of Congress, Prints and Photographs Division (LCCN 2005694947)*

predict shipping schedules; in this way, fields and factories alike came to depend on industrial concepts of time.[12]

Industrialized agriculture required new and expanded technologies of food preservation. Salting and drying meats had been practiced for centuries before Nicholas Appert developed canning in France at the outset of the nineteenth century, when supplies of sugar and salt, ingredients traditionally used to preserve foods, were cut off by the Napoleonic Wars. Canning allowed food processors to store their goods for months or even years, and offered packaging for easily transporting and stocking them.[13]

American tomato production became increasingly mechanized after 1870. Prior to that time, canning was done by hand, with recent immigrants and African Americans performing the bulk of this labor and working for low wages. Over the following half century, tomato producers introduced a series of innovations to distance the process from human hands. These included machines that made and capped the tin cans and steam retorts that functioned like pressure cookers to reduce cooking time and more reliably kill bacteria. By 1920, the largest companies had automated most of the tomato-canning process and boasted that "'no human hands' had touched their products."[14]

As production methods evolved between 1850 and 1950, so too did transportation systems and outlets for consumption. Reliable rail transit (including refrigerated cars to store meats) nationalized food systems—a trend strengthened with the construction of highways and proliferation of trucks, as well as iron ships and shipping containers that allowed for reliable transoceanic imports.[15]

Human labor was and is an important part of industrialized food systems across the world, with wage or forced labor playing roles in the cultivation, harvesting, and processing of foods ranging from oranges to bananas to meat. Workers involved in food systems fall into contested categories of "skilled" and "unskilled." These include seasonal field workers paid low wages, exposed to chemical and environmental hazards, and at risk of replacement. Workers also include the chemists and engineers who serve to coordinate systems of fertilization, pest control, irrigation, and crop yield.[16]

The skills and other requirements of agricultural work have evolved over time, as has work throughout industrialized society. Prior to the mass production of meat in slaughterhouses, skilled butchers who apprenticed for years served local markets. The increase in the world of edible goods allowed consumers to bypass the skilled butcher and local farmers in favor of possibly cheaper outlets for more-distant foods. After World War II, an increasing number of American consumers could find their groceries at supermarkets, a trend that swept through much of the industrialized world by the end of the century. Centralization of food production—and the ability to ship foods great distances—reorganized labor, so that most human labor involved in food systems rendered services such as stocking packages and exchanging money for mass-produced foods often harvested hundreds or thousands of miles away.[17]

Many of the foods found in markets required packaging more varied than the cans and bottles of the prewar era. By the 1960s, "TV dinners," apportioned in metal trays and foil and encased in cardboard outer packaging, became staples of US consumption. In the half century since their introduction, plastic trays, caps, and wraps have become recognizable parts of the packaging of many foods, including those marketed as convenience foods or suitable for packing in children's lunches. The proliferation of packaging has increased the amount of solid wastes produced and landfilled, adding economic and environmental burdens to municipalities. Industrial systems produced goods marketed as individual conveniences to consumers, even as they shifted responsibility for disposing of packaging from producers to consumers and municipalities.[18]

Food production was accelerated by the industrial reorganization of water supplies to ensure steady sources of irrigation. Dams made of steel and concrete provided this water; they also appealed to governments because they provided hydroelectric power and thousands of construction jobs. By the mid-twentieth century, governments in the United States, the Soviet Union, India, Egypt, and elsewhere were in the business of building dams. These dramatic interruptions of waterways produced immense bounty in agricultural products, drinking water, and hydropower. Over time, they also revealed the limits of extracting value from river systems as ecological damage and declining farm productivity raised questions about the long-term sustainability of societies heavily dependent on dams.[19]

States in nations around the world also regulated land use and applications of fertilizer and pesticides and subsidized cultivation of particular crops. Whether nations had economies defined as command-control or free market, state regulatory instruments helped industrialize rural land, waterways, and food production.[20]

Dedication of resources to maximize agricultural yields became known as the Green Revolution. During the second half of the twentieth century, states in giant nations such as India, the People's Republic of China, and the Soviet Union, as well as in smaller nations such as Sri Lanka, pursued policies to increase yields of grains and livestock. State-sponsored efforts to industrialize food systems involved coordination with universities and

Irrigated forest plantations, like this one in Punjab, India, in the mid-twentieth century, showed how the industrial reorganization of water for irrigation could also alter forests and agricultural landscapes. They also challenge common definitions of technology. As the linear, planned layout of this forest suggests, experts treated trees as technologies to maximize growth, yields, and therefore profit. Yet, fundamentally, these forests still relied on water. Irrigated forest plantations thus illustrate the idea of envirotechnical systems. *G. D. Kitchingman, "The Punjab Irrigated Plantations," Empire Forestry Service 23, no. 2 (December 1944) via Wikimedia Commons*

corporations to innovate new techniques, chemicals, machinery, and even genetic engineering.[21]

The African experience with maize in the second half of the twentieth century represents part of the Green Revolution that transformed industrialized agriculture worldwide. Maize came from the Americas and became dominant in African fields in the twentieth century. Though not native to the continent, the grain grew fast and required less labor than other grains. Given political instability and fears of drought in nations like Ethio-

Engineer J. A. L. Horn took this photograph of Japanese rice terraces in 1935. Rice production, including "wet field" cultivation, has a long history in Japan. Rice terraces suggest a highly managed approach to raising crops that materializes in the landscape itself. Nonetheless, rice raised in this manner still depended on water. *National Museum of Denmark via Wikimedia Commons*

pia, corn became popular with farmers seeking reliable crops. Breakthroughs in cultivating urbanized plants, with the aid of synthetic fertilizers and pesticides, to produce greater yields of grains (which in turn fed humans and livestock) transformed food systems in Africa, the Americas, Asia, Europe, and Australia, and those of several island societies.[22]

A crucial dimension to this transformation was the human alteration of the nitrogen cycle (the process by which nitrogen is converted into multiple chemicals as it circulates through the atmospheric, terrestrial, and aquatic ecosystems), initially in the nineteenth century, and then with growing intensity in the twentieth century. Instead of relying on the natural ecosystem services that plants and animals perform in returning nitrogen to the soil, chemists harnessed nitrogen on a far wider scale to maximize the land's productivity. Similar manipulations of phosphorous led humans to move from harvesting the mineral from bird excrement to mining and distributing it in concentration. This engineering achievement increased crop yields, along with the productivity and affordability of foods, allowing the worldwide human population to grow from just over two billion in 1930 to over seven billion in 2015. It also upset the balance of aquatic ecosystems, producing algal blooms that choked oxygen from reaching indigenous organisms, creating dead zones in oceans and seas, and exacerbating environmental consequences for industrialized food production.[23]

## Industrializing Agricultural Disasters

While human manipulations of nature have shaped the food systems we have depended on, they have also produced long- and short-term consequences for air, land, water, other species, and human health. By 1000 BCE, deforestation carried out in the Chinese uplands to clear land for agriculture had resulted in centuries of silting and flooding. In the twentieth century, industrialized agriculture accelerated the speed and magnitude of environmental consequences. Donald Worster argues that an industrial capitalist treatment of the land for harvesting monocultural crops produced the soil erosion in the United States' Great Plains that became known as the Dust Bowl (when soil erosion affected more than 100 million acres in Texas, Oklahoma, New Mexico, Colorado, and Kansas, displacing tens of thousands of farmers during the 1930s). The devastation of the region's farms during the Dust Bowl was—far from being a "natural disaster"—a human-engineered disaster. Dams allowed the expansion of agricultural production on previously arid lands; yet such transformations meant that the flora and fauna of affected lands "went through an upheaval comparable only to the cataclysmic postglacial extinctions."[24]

Industrialization transformed lands used for intensive growing of plants and animals in rural areas. Demand for grain and meat in urban centers meant that rural landscapes were planted in ever-increasing amounts of corn, wheat, rice, and soybeans, and used for ever-larger livestock operations preparing cattle, hogs, chickens, sheep, and other animals for slaughter in centralized urban meatpacking facilities.

Industrial meat production centralized slaughtering and processing; in the United States, these activities were concentrated in Chicago, where the meatpackers dumped their wastes into the South Fork of the South Branch of the Chicago River. A small tributary with a weak current, the waterway was immediately overwhelmed with meatpacking plant wastes in the 1850s; by the 1880s, it earned the nickname "Bubbly Creek" because of the methane bubbling from decayed organic matter. Bubbly Creek emitted odors that nearby residents complained about. The volume of solid waste entering the stream by 1911 produced a skin atop the water's surface that chickens and even humans could stand on.

The consequences of industrial meat production were conspicuous, especially after their vivid description in *The Jungle*, in which author Upton Sinclair described Bubbly Creek as a "great open sewer."[25] Environmental damage from other food production methods could be even greater, if less obvious. Soil erosion had consequences for the humans and animals who lived on affected lands. The hydroelectric dams that governments in the United States, the Soviet Union, China, and India constructed between 1920 and 1970 reshaped watersheds, lands, and the fortunes of plants and animals within ecosystems. Despite the addition of "salmon ladders," structures designed to allow migrating fish to pass over or around dams in the US Pacific Northwest, salmon populations in the Columbia River plummeted. In response, commercial fisheries developed salmon farms, which produced commodity fish while creating new challenges relating to disease and genetic diversity. The engineering of waterways represented a disaster for wild salmon that commercial food producers mitigated with domesticated fish.[26]

An enduring theme involves the technologies humans develop to control nature, and the intended and unintended consequences that result from the applications of these technologies. Chemical manipulation of ecosystems had ramifications. Beyond the aforementioned interruption of the nitrogen cycle, monoculture production necessitated eradicating unwanted plant species (classified as weeds) from fields, a process often involving the use of chemical herbicides. Aside from effects on animals that ingested the herbicides and on the groundwater beneath the fields, herbicide use produced homogenous yields of a single crop. Though a boon for industrialized agriculture and its quest for maximizing value extracted from the land, successful monoculture bore the unintended consequence of producing abundant food for unwanted organisms classified as pestilence. Species that could eat acre after acre of crops could multiply exponentially and devastate harvests. A popular solution to this problem was to apply more chemical poisons, classified as insecticides, to fields. Insecticides also had potential consequences for human and livestock health, necessitating washing of harvested foods. As with herbicides, insecticides exposed agricultural labor to poisons and seeped into groundwater, poisoning drinking water for humans and animals.[27]

Industrial research and development during the twentieth century produced new herbicides and insecticides focused on particular threats. The insecticide dichlorodiphenyltrichloroethane (DDT), developed for military uses during World War II, saw widespread application in the postwar era to "protect" human and livestock health. This expanded use exacerbated observed hazards of the chemical as a toxin to fish, birds, and humans as it accumulated in the tissues of plants and animals and caused illness and death. Rachel Carson, alarmed by the consequences, published *Silent Spring* in 1962. The book's title referred to the absence of birdsong due to the devastating impact of chemical poisons on bird populations. *Silent Spring* altered public discourse on the merits and perils of chemical innovation. Results have included a federal ban on DDT use in the United States and sustained public and regulatory concerns about the effects of agriculture-related chemicals on human and environmental health.[28]

Food systems raise broad issues of continuity and change within the history of technology. Changing technologies in industrial society have led to shifts in the ways by which people acquire and prepare food. Stoves and ovens powered by electricity and natural gas became standard appliances in American homes in the second half of the twentieth century, increasing food preparers' control of temperature. The microwave oven became a mass consumer amenity after 1970; its history shows the complex interactions between technology and sensory history. Microwaves increase convenience; they also produce tastes and textures different from those electric or gas stoves produce.

Increased convenience in domestic food preparation appliances shaped the production and packaging of foods. The proliferation of ice boxes in the early twentieth century expanded the reach of American dairies as delivery of milk in reusable glass bottles became commonplace. Prepackaged foods could save consumers time at the market, reshape where and what kinds of foods were available, and expand the reach of food producers, from local to regional and global scopes.[29]

Industrialization's reshaping of a host of institutions, from schools to hospitals to road systems, has produced concomitant changes in food systems. Transportation systems—including railroads with refrigerated cars, interstate highways, ships, and airplanes—have allowed food producers to "annihilate geography." By the turn of the twentieth century, Chicago could proclaim itself "Hog Butcher to the World" because pork products shipped from there across the United States.[30]

In the twenty-first century, transportation networks allow diners in New York City sushi restaurants to eat fish caught in Japan, butchers in London to sell lamb from New Zealand, and grocery customers in Minneapolis (USA) to eat blueberries grown in Chile. The rise of modern domestic kitchens after

World War II led to packaged prepared foods. The microwave oven allowed mass production of new foods, such as bags of microwave popcorn.[31]

Industrialization also shaped foods people could eat in retail outlets and other institutions. Interstate highway construction led to the rise of fast food in the United States, with White Castle's proliferation in the 1920s anticipating a plethora of restaurants offering affordable, elaborately packaged and highly processed meals suitable for eating in a car. McDonald's, Pizza Hut, Taco Bell, and dozens of other chains use this model in thousands of restaurants around the world. The growth of schools, hospitals, and corporations created markets for cafeterias that purchased and produced foods in bulk, at times valuing preservation over taste. Packaged, processed foods became commonplace in these institutions, joining fast foods as time- and money-saving innovations in the food humans ate.[32]

The large technological system of industrialized food production required further reorganization of land, capital, labor, and technology as inputs. Monoculture produces higher yields; it also leaves crops vulnerable to predators adapted to eating particular crops. Pestilence produces blight and economic loss. Industrial food systems' adaptations in response to pestilence include the use of chemical pesticides and herbicides. One of the reasons for *Silent Spring*'s impact on the public in 1962 is that readers linked the chemical innovations to harm to humans, and this narrative has informed discussion of chemicals in food production for more than half a century.

As Upton Sinclair wryly noted about the reception of his 1904 exposé, *The Jungle*, public concern about food systems focused on the quality of the food produced rather than the working conditions of the people employed to produce it. This pattern of regulation, focused largely on the food rather than the workers, continued throughout the twentieth century. Industrialized foods have led to conflicts between regulators and consumers over food safety and quality, with laypeople struggling to combat technical experts' definitions and measurements of safety. Examples include battles over the quality of Mexico City (Mexico) sausage at the turn of the twentieth century, and between consumer advocates and the FDA over peanut butter in the 1960s and 1970s.[33]

Industrialized food systems transcend political boundaries; Dole Fruit and C&H's interest in the Pacific helped shape the United States' interest in Hawaii as a territory in the 1890s and the islands' subsequent admittance as a state, in 1959. One of the results of the 1994 North American Free Trade Agreement was an expansion over the next decade of tomato crops grown in Mexico for consumption in the United States.

Transnational activities complicate the regulation of pesticide and herbicide applications, posing health risks to agricultural workers as several carcinogenic and endocrine disruptions (discussed in chapter 5) are associated with agricultural chemicals. Susanna Rankin Bohme points to lower male fertility among Hawaiian and Central American workers exposed to the pesticide dibromochloropropane (DBCP). These hazards were results of humans seeking ever-greater bounty from the land, and they raised worries by the 1970s about the costs to environmental and human health that such efforts exacted. Those worries would intensify by the end of the twentieth century.[34]

## Genetic Engineering and Food Systems

Recent concern about food systems centers on the genetic engineering of organisms. Although the history of genetically engineering commodity foods begins in the 1980s, such work has important continuities with earlier efforts to breed plant and animal species. In one of the landmark books in the history of technology and the environment, *Industrializing Organisms: Introducing Evolutionary History* (2004), the contributing authors discuss several of these efforts in the nineteenth and twentieth centuries. More than a decade after its publication, the authors' analyses of nineteenth-century manipulation of sugar, wheat, and cows, as well as of twentieth-century chicken and hog production, remain relevant to our understanding of industrialized food systems in general, and the historic precedents for genetically engineered organisms that shape the twenty-first-century food supply.[35]

The precedent established, genetic engineering (GE), is also dramatically different from past manipulations. Writing in *The Illusory Boundary*, Edmund Russell argues that "genetic engineering offers a faster, more precise way of doing what breeders have long done. But the novelty of genetic engineering lies in its ability to move genes across wildly divergent taxonomic groups, such as between plants and animals. Scientists implant firefly genes in tobacco plants and frogs to make them glow, and rice plants in Missouri (USA) manufacture human proteins courtesy of imported human genes."[36] Russell concludes, "Opponents of genetic engineering usually portray it as something radically new, which it is. Never before have we been able to move genes between plants and animals. Proponents of genetic engineering, on the other hand, portray it as but the latest phase in the production of biotechnologies, which it also is. For all of us who think history has something useful to say about the present and future, it is essential that we recognize both the disjunction and the continuity between past and present."[37]

In the twenty-first century, the most widespread use of genetic engineering involves corn and soy crops, which serve as food supplies for livestock. About 100 million acres of farmland in the United States were planted with GE crops in the year 2000. According to one estimate in 2001, more than two-thirds of the food in America's supermarkets involved genetically engineered organisms in some proportion. As Ann Vileisis notes, "Corn, in particular, was engineered to include the bacterium *Bacillus thuringiensis*, which produced its own insecticide, Bt. Subsequent studies revealed that Bt expressed in the pollen of GE corn was toxic to beneficial insects, which raised the broader ethical and policy question of why ecological studies had not been conducted *before* the GE crop was released for use on millions of acres of farmland."[38]

Economics and taste have influenced the concentration on these crops, which have become vital to both processed foods and industrialized meat production. Corn's application in high-fructose corn syrup, which became an economically viable replacement for cane sugar by 1980, has made corn a staple of processed foods from the obvious (soda, candy) to the subtle (bread, pasta sauces). Surveys of convenience stores in the United States reveal a majority of the edible products for sale include corn in some way. Genetically engineering the crop for maximum yield makes economic sense, even as it further intensifies a mono-culture that requires interference in the nitrogen cycle and demands killing or repelling unwanted organisms in the fields.

Meat-based diets rely increasingly on genetically engineered crops. In the United States, increased corn production between 1970 and the end of the century meant that cattle's diets shifted from grasses to corn, producing animals that reach market weight faster and have fattier tissue in the harvested meat. With less protein and more fat, the American hamburger in 2015 tastes and feels substantially different from its 1970 predecessor. Hogs, too, consume more corn and soy because of the economics of those grains, with similar outcomes in the meat thus produced.

The continuing debates surrounding GE organisms reveal the complex interactions of technology and nature, and the potential consequences to the food system, to human bodies, and to ecosystems. The foods that people living in twenty-first-century industrial society eat are the dynamic products of nature and technology, evolution and history. Such a statement may seem obvious, with the proliferation of highly processed foods such as breakfast cereal, bread, candy, ketchup, and soda, but it is also true of corn, tomatoes, carrots, rice, beef, pork, and chicken. Russell reminds us that cheese production is a biotechnological process, as is the brewing or distillation of alcoholic beverages such as beer, wine, and spirits.[39]

Modern food systems reflect such enmeshed relationships between technology and the environment that reactions against them reveal industrialization's depth and complexity. The use of fertilizers, pesticides, hybrid strains designed for maximizing caloric content, chemical preservation, and other processing has raised concerns over the quality of the food supply and its effects on ecosystems and human health. Groundwater pollution and cancer clusters in agricultural regions joined alarm over increases in obesity, diabetes, and cardiovascular disease in much of the industrialized world by the end of the twentieth century. Since the late 1960s, organic and locavore food systems have grown in reaction to industrialized agribusiness. These developments, however, have resulted not in preindustrial modes of food production, but rather alternative industrial systems that differ from the mainstream systems more in degree than in kind. By 2010, the chain supermarkets Whole Foods and Walmart had developed systems of distributing both organically produced and locally grown foods to the same store shelves that stocked globalized processed foods.[40]

This chapter offers a few of the ways in which historians have explored how the food humans eat represent past and present mediations between technology and the environment. If the packaged candies and sodas on supermarket shelves represent the most obviously artificial engineering of raw materials into edibles, they sit on a continuum with the meats, fruits, vegetables, grains, and beverages that have structured industrial society's diets. Historically, these foods and the systems that deliver them have allowed humans to multiply and thrive. But they have also resulted in serious consequences for ecosystems, biodiversity, and human health.

Attempts to create more-sustainable food systems in the future must take this history into account. "If many billions of people are to live on Earth peacefully and equitably in thriving economies," Hugh Gorman argues, "industrialized societies have no choice but to construct a guide that places ethical and practical boundaries on human interactions with the planet." Gorman is one of the school of historians of technology and the environment whose work shows how central the food systems developed to serve humanity have become in producing this vexing problem.[41]

# Chapter Two

# Industrialization

The previous chapter shows how industrialization played a key role in the remaking of food and food systems in many parts of the world. We now turn to the history of industrialization more broadly, examining it through the lenses of technology and the environment. Doing so offers new insights and, ultimately, new interpretations of this important era in human and planetary history. For one, it shows how the dynamics among technological, social, and environmental change in industrial development—and far beyond—are more complex than one process simply driving the others. Contrary to many popular histories, the steam engine and cotton gin did not simply "cause" "the Industrial Revolution." We therefore need more-sophisticated ways to discuss and represent these dynamics.[1]

In this chapter, we make three arguments. First, industrialization not only shaped, but was also shaped by, nonhuman nature. Second, it required and produced not only new social relations but also new relationships between people and the natural world. Third and relatedly, industrialization both reflected and depended on major transformations in human-natural dynamics in terms of time and space.

## What Is Industrialization?

Before turning to these arguments, we should ask, What *is* industrialization? Most broadly, industrialization describes the shift from an agrarian society, where economic livelihood and wealth are derived from the land, agriculture, and artisanal production, to an industrial, increasingly urban society centered on machine-based manufacturing largely dependent on fossil fuels. "The Industrial Revolution" generally refers to industrialization in parts of Western Europe and North America, especially Britain, between the mid-eighteenth and mid-nineteenth centuries. It centered on significant developments in the steam engine, coal mining, and textile manufacturing. Although many histories have focused on these technical achievements, industrialization also depended on raw materials such as cotton and sugar, labor, and export markets in European colonies. These collective changes are sometimes called the *first* Industrial Revolution. The *second* Industrial Revolution, between the mid-nineteenth and early twentieth centuries, was rooted in large-scale technological systems such as railroads, electrification, and eventually the internal combustion engine dependent on oil.[2]

Historians, sociologists, economists, and other scholars have studied industrialization for as long as these historical processes have taken place. Scholarship has often focused on questions of class and labor, particularly since publication of the influential writings of German scholars and social critics Karl Marx (1818–83) and Friedrich Engels (1820–95).[3] Later historical research explored how gender, race, and empire were also crucial to in-

This engraving, "Cottonopolis," by Edward Goodall (1795–1870), shows Manchester, England, the metropolitan hub of the British Industrial Revolution. Manchester was nicknamed Cottonopolis because it was the center of cotton processing and textile production. The engraving suggests the environmental impact of industrial mills and factories. It also contrasts the dirty city with the clean environment of the surrounding countryside. In fact, many hinterlands, whether domestic or colonial, were linked to and depended on cities. *Wikimedia Commons*

dustrialization. Industrialization amplified a gendered division of labor, enslaved people, and produced key resources like cotton; and colonies provided many of the raw materials for industrial production, as well as markets for industrial goods.[4] Historians have also questioned whether the social, economic, and technological changes between roughly 1750 and 1850 should even be described as a revolution—as a radical, rapid break with the past. English economic historian Arnold Toynbee popularized the term *Industrial Revolution* in the late nineteenth century; while the phrase may reflect the assumptions of his own time, it may not accurately characterize actual historical change.[5]

## Energy and Industrialization

Industrialization certainly affected the natural world in many ways and the Industrial Revolution's environmental "impacts" are one of the better-known aspects of its history: smoke, soot, deforestation, polluted rivers, even the evolutionary selection of dark-colored moths near England's famous mills as factory particulates darkened industrial cities.[6] Both novelists and non-fiction writers of the mid-nineteenth to early twentieth century, such as Charles Dickens, Elizabeth Gaskell, and Upton Sinclair, described to great effect in their writings the environmental and human health problems associated with industrialization.[7] However, for some contemporaries, these were not problems but demonstrations of economic prosperity. In the early twentieth century, Yawata (Japan's first large steel town) celebrated smoke and grime as signs of wealth, modernity, and progress. As its civic anthem proclaimed, "Billows of smoke filling the sky / Our steel plant, a grandeur unmatched: / Yawata, O Yawata, our city!"[8]

Smoke and soot suggest how energy sources, and the environmental consequences of new forms of energy and new rates of energy extraction, were critical dimensions of industrialization. Historians have traced the growing use of fossil fuels in industrial production and even as the basis of the modern political economy.[9] Although coal had been used to heat homes in London since at least the seventeenth century, consumption began to rise dramatically in the mid-eighteenth century.[10] Not only did the demand for coal increase, but the energy source began to be used in new ways. In particular, coal fueled repetitive, linear motion to do work.

Rotational motion driving pumps had a long history, dating back to ancient Egypt. Cams allowed the transformation of rotational into linear motion, which is more efficient. These technologies also have a long history, dating to at least the third century BCE in both China and Greece.[11] Widespread deforestation in Europe by the eighteenth century, however, helped encourage the exploitation of coal reserves. Significant energy sources were needed to fuel the conversion from rotational to linear motion in ways that would substantially increase the work that such motion could perform. A Frenchman proposed the first steam engine in 1689, but it was never built or patented. Blacksmith Thomas Newcomen designed and built the earliest steam engine in 1712, but it was inefficient. James Watt developed a more efficient steam engine in 1769. He continued to make improvements, resulting in his famous 1783 model. A range of manufacturers adopted Watt's enhanced model over the following two decades, including those in the mining, iron, textiles, paper, and flour industries. This extended period of design and development—from the late seventeenth through the late eighteenth century—illustrates why some scholars have questioned the supposedly revolutionary character of the so-called Industrial Revolution. Nonetheless, this series of improvements eventually resulted in the ability to turn heat into work more efficiently and specifically into mechanical motion that would drive machines in entire industries. And these heat engines were fueled (literally) by millions of years' worth of photosynthesis—in other words, coal.[12]

## *Fossil Fuels*

Coal consumption increased during the eighteenth century, but it skyrocketed over the nineteenth: from 10 million metric tons in 1800 to 1,000 million metric tons a century later—a hundredfold rise in as many years.[13] Coal use continued to rise during the twentieth century, but at a much slower pace, growing only fivefold during those hundred years. Total global energy use increased eightyfold from 1800 to 2000.[14] In short, these trends demonstrated a pattern of energy intensification.

By the twentieth century, such intensification and the primary dependency on fossil fuels marked a radical transformation in human interactions with the environment. The adoption of coal, however, was not as easy or straightforward as these striking figures may at first suggest. Not all coal has the same qualities and properties. Industrialists and domestic consumers therefore needed to learn, for example, how to use carbon-dense anthracite coal (also known as hard coal) effectively.[15]

The industrial use of petroleum, one of the bases of the second Industrial Revolution around the turn of the twentieth century, is more recent, but it followed a similar pattern of rapidly growing use, eventually surpassing coal in importance. Unlike coal, oil constituted no share of energy consumption in 1800. In the mid-nineteenth century, oil drilling—initially in Pennsylvania (USA) and then soon at many other sites—began to transform the relative share that oil represented of fossil fuel use. In 1900, 20 million metric tons of oil were consumed, as compared to 1,000 million metric tons of coal. By the turn of the twenty-first century, though, consumers used 3,000 million metric tons of oil. Coal and oil demonstrate two key shifts, then: first, significant increases in total energy use from the eighteenth through the twentieth century; and second, the nearly complete replacement of biomass energy sources with fossil fuels.[16]

Historians have thus traced the roots of fossil fuel dependency—now blamed for climate change—to the eighteenth century, examined how energy dependence accelerated over the nineteenth century and exploded after World War II, and shown how fossil fuel use became a global phenomenon in the twentieth and early twenty-first centuries. Collectively, these trends explain how the amount of carbon dioxide in the atmosphere has increased since 1750.[17] Some scientists have argued that human effects on the planetary atmosphere are so great that we need to designate a new geologic epoch: the Anthropocene.[18] Industrial modes of production thus radically transformed geologic processes and scales. Although intensification therefore characterized fossil fuel consumption during the final quarter of the second millennium, such intensification was also marked—and remains marked—by major inequalities. Not all people have consumed an equal share of fossil fuels. As such, they have not all born equal responsibility for "global" climate change.[19]

As J. R. McNeill, Vaclav Smil, and others have shown, the volume and pace of energy consumption rose dramatically in recent human history, but extracting, processing, harnessing, and distributing such energy have also had significant socio-environmental repercussions.[20] Underground coal mines required miners, foremen, and owners to develop new ways of knowing, visualizing, and managing labyrinthine networks of coal seams and tunnels.[21] By necessity, miners had intimate knowledge of and interactions with subterranean environments.[22] Yet mine collapse threatened workers with disease, injury, and even death, while tailings and slag contaminated water, soil, and also human bodies.[23] Open-pit coal mining, rather than underground mining in tunnels, increased the scale of effects—altering surface topographies, watersheds, even cities.[24] As oil followed water downstream, oil seepage and blowouts merged hydrologic and energy systems in new ways.[25] Spurred by the expansion of the natural gas industry in the late twentieth and early twenty-first centuries, fossil fuel extraction through hydraulic fracturing ("hydrofracking") has caused anthropogenic earthquakes, showing the ability of people to affect geologic processes such as seismic activity.[26] Harnessing rivers through large hydroelectric projects dramatically transformed many watersheds over the twentieth century, from the Amazon River (which includes parts of Brazil, Colombia, and Peru) to the Yangtze (or Chang, in China).[27]

Meeting rising energy demands has had additional unexpected effects. At times, "old" and "new" energy sources coexisted and worked in concert. For instance, US industrialization in the nineteenth century actually created greater demand for horses. In the mid-nineteenth century, newly introduced "steam" ships limited costly (and sometimes politically tricky) coal refueling by also using ocean currents and sails.[28] As these examples suggest, "modern" technologies simultaneously depend on nonhuman resources, processes, and species. In this case, horses and steam engines actually coexisted and complemented each other. There was no tidy, linear, complete shift from one form of transportation or work to another. Such cases have pushed envirotech scholars to begin analyzing diverse energy systems as envirotechnical systems to foreground this entanglement between the environmental and the technological.

## Industrialized Nature
### Other Inputs

*Intensification* also described the production and extraction of other "natural resources" for industrial production. The term in fact reveals an industrial mindset: the transformation of natural entities into discrete, extractable, valued commodified goods. Forests served, of course, many purposes—from the ecological to the aesthetic—but trees fed paper mills, and forges were fueled by charcoal (carbon residue produced by heating wood without oxygen).[29] Plantation agriculture in the US South and South Asia, shouldered by laborers under violent and exploitive systems, produced large volumes of cotton that were woven into textiles in European and New England mills.[30] Other resources became parts of the machines themselves. Bison hides, for instance, were turned into belts that literally turned the wheels of factory production.[31] In general, the pace and output of industrial manufacturing utilized vast quantities of raw materials at unprecedented rates with major consequences for the people and landscapes involved in producing, extracting, and processing those resources.

This photograph, taken from the book *Indian Cotton* (1915), shows the "Rajah" plow, developed for raising cotton in early twentieth-century India. More common among "ordinary smallholders," this plow serves as an important reminder that many resources were extracted from colonies and distant locales. The photo also visually represents how the "technology" of the plow included people and animals, not just a material artifact. *Wikimedia Commons*

These industrial methods fostered, even facilitated, still more extraction. For example, railroads moved grain, meat, hides, and timber from the US West to urban, industrial centers like Chicago—and quickly.[32] Meanwhile, growing demand for whale oil for industrial and domestic illumination pushed the expansion of whaling in both the Pacific and the Atlantic Ocean during the nineteenth century.[33] Larger ships, fueled by the steam engine and with processing machines on board, then further accelerated whaling at industrial scales.[34] As a result, the production of whale oil through industrialized whaling began to outpace the biological reproduction of whale populations.[35] In other words, natural resources around the globe, often under-girded by coercive forms of slave or colonial labor, fueled the technological systems of industrial production, while such technologies enabled faster and still-larger harvests of those "resources." Industrial technologies thus had significant implications for valued species and ecosystems. Such patterns occurred on a global scale as railroads in the colonies and steamships connecting empire and metropole facilitated this process for European industrial centers.

As a result, species and landscapes changed at both micro and macro scales. Industrialization transformed some ecosystems like forests and watersheds to better suit industrial production. Forestry science helped maximize timber production, first in Germany and then throughout the world as the German model shaped many national and colonial forest services. By doing so, professional forestry experts hoped that natural resources and environmental systems would become efficient tools; in this sense, natural entities were also understood and functioned as technological.[36] Engineers dredged, harnessed, and redesigned rivers in an attempt to control flooding, increase energy production, expand agricultural production, and provide steady supplies of water to municipalities, farms, factories, and eventually nuclear reactors.[37] Many of these environments underwent dramatic transformations as scientists, engineers, and other techno-scientific experts simplified, commodified, and literally remade nonhuman nature. Canals expanded and deepened some existing natural waterways. Similarly, some irrigation networks in arid regions built on existing water flows and topography, even as they modified them.[38] Other canals connected bodies of water that had never been linked before. Compared to their predecessors, modern, industrial canals vastly expanded the size and number of watercraft that could travel between bodies of water, as the histories of the Suez (Egypt, 1869) and Panama (Panama, 1914) Canals suggest.[39]

As the previous chapter shows, early forms of what we now call industrial agriculture involved imported species and breeding organisms better suited to industrial methods of production and harvesting—again changing ecosystems at multiple scales, from the organism to the entire landscape.[40] Monoculture itself was an industrialization of the landscape, as raising vast tracts of a single crop entailed the drastic simplification of environments in an attempt to maximize food production. In the late nineteenth century, cattle, wheat, and corn transformed enormous expanses of land in Argentina and the US West. In later decades, ranching and soybean and sugar beet farming remade Brazilian agriculture. Favoring a limited number of certain crops (and therefore species) frequently entailed the simultaneous extinction or near extinction of others. The importation of the Holstein cattle breed (known as the highest-production dairy animals) into Central Africa after World War II increased the availability of milk and other dairy products for many families and communities, but interbreeding now threatens the genetic diversity of Ankole-Watusi cattle, a species that resulted from interbreeding between African and Indian cattle two thousand years ago.

Before these transformations took place, boosters conveniently saw many of these landscapes as empty, "virgin" lands. Such perspectives helped legitimate conquest, expropriation, and settlement. But scholars have shown instead that it was colonialism and empire that emptied these landscapes, as native peoples, ethnic minorities, and other marginalized groups were often displaced from and dispossessed of their historic territories.[41] During the Qing dynasty in China, rulers expanded their empire by conquering nomadic peoples and encouraging them to immigrate to promote colonization of oases in the desert.[42] Overall, energy sources and natural resources were critical inputs to industrial production. Extracting, producing, transporting, and processing these inputs all had major environmental consequences.

## Outputs

We can also consider the other end of industrial production, namely, the waste resulting from manufacturing. Indeed, pollution associated with industrial production has been well documented, at times by contemporaries.[43] Some early scholarship by historians of the environment, technology, and cities studied diverse pollutants—from the organic dregs of tanneries, abattoirs (slaughterhouses), and sewers to new, synthetic products. Historian of technology and the environment Joel Tarr has described how many industrialists searched for "the ultimate sink": a disposal site where effluents and contaminants could be discharged without considerable cost, resistance, or hazards that might then spur investigation. In other words, they believed waste could be put "out of sight, out of mind" and had confidence that the "ultimate solution to pollution is dilution," whether by air or by water. At other times, waste was transported and dumped far from the site of production, conveniently reducing exposure at industrial sites but increasing exposure "elsewhere."[44]

In studying the history of industrial pollution, some historians have focused on an industrial heartland, such as Germany's Ruhr Valley or the US city of Pittsburgh, and how the landscape and waterscape of the surrounding countryside were affected.[45] Similar concerns have arisen recently regarding China's rapid industrialization in the late twentieth and early twenty-first centuries. Different parts of the environment—for instance, air or water—have been contaminated over time across the world. Particularly in the nineteenth and early twentieth centuries, industrialization brought about increasing pollution of rivers, lakes, and oceans with human and animal sewage, the waste of tanneries and early mills, and eventually inorganic chemicals and radioactive waste, along with significant air pollution.[46] Public recognition of pollution was, however, often slow, and actions to address the problem even slower. Air pollution in Britain, for instance, emerged gradually as a social and political problem during the second half of the nineteenth century, but remained a major issue for decades.[47]

At times, some forms of pollution crossed political borders, exposing tensions between ecological and political boundaries. In the early twentieth century, smoke and particulates produced by smelters—metallurgical plants that extracted silver, iron, or other metal from ore—did not respect the border between the United States and Canada. Similarly, acid rain in Europe created novel political challenges and solutions for a relatively small continent containing many nation-states. Air pollutants did not follow even property lines; the wind sent particulates

This photograph, taken by John L. Alexandrowicz in May 1973, documents the dumping of slag (stony by-products of refined or smelted ore) at a steel plant in Pittsburgh, Pennsylvania. Still warm when it hit the water, it produced a haze that lingered over downtown (shown in the photo's background). This is just one example of pollution resulting from industrial manufacturing and processing. *US National Archives and Records Administration (557232)*

This 1971 stamp from the Soviet Union is an idealized representation of agriculture and Communist labor. Nonetheless, its imagery of agricultural machinery, mechanized harvest, and electricity suggests a commitment to high-modernist agriculture and industrial production. *Wikimedia Commons*

whichever way it blew. Building taller smokestacks often sent pollutants greater distances. Doing so could conveniently diffuse causality and blame.[48] Technological artifacts could thus facilitate the crossing of borders. Conversely, unpolluted air was especially valued, though the practices of one country might affect the health of another country's air.[49] Furthermore, despite seemingly dramatic differences among political systems, distinct political ideologies frequently shared common environmental impacts. The supposedly opposing political systems of capitalism and communism in fact shared many commonalities, including a commitment to high-modernist industrial production with its attendant environmental consequences.[50]

Both bottom-up and top-down reform movements in the nineteenth and twentieth centuries sought to address pollution, albeit typically in different ways.[51] Building "modern" infrastructure like water and sewer systems reflected an attempt to manage greater concentrations of people, production, and pollutants in the industrial age, usually through experts, bureaucratic institutions, and large-scale infrastructure.[52] Yet controversies over the extent—and sometimes even the existence—of pollution and technical solutions, and over who should pay for these fixes, have marked and often delayed regulation since at least the nineteenth century.[53]

Environmental and social historians have also shown how these hopeful assumptions that experts and technology would "fix" the problem did not always prove true. Technical elites may have hoped that their technological systems were distinct from environmental systems, but accidents and crises exposed the fallacy of this thinking. Even if they were successful in moving pollutants, contaminants, and waste "elsewhere," their "elsewhere" was someone else's "here." As the chapters on discards and the body reveal in more depth, technological systems have at times simply displaced the environmental consequences of industrialization from one community to another.[54]

## Body-Environment

Pollutants affect not just nonhuman species and ecosystems, but also people, thereby blurring the boundary of the human body and the environment—a central theme of chapter 5, on the body. Moreover, the socio-environmental consequences of discarded industrial materials (discussed in greater detail in the following chapter) highlight a related key point. Both scholars and some contemporaries have emphasized how people's own bodies and ultimately their health were detrimentally affected by industrial modes of labor and industrial landscapes.

In an influential article published in the mid-1990s, environmental and legal historian Arthur McEvoy argued that people's bodies should also be seen as biological components of industrial systems. The "nature" of industrialization typically implies natural resources such as coal, cotton, or timber, or landscape features such as rivers, forests, or the atmosphere. McEvoy instead pushed environmental historians to consider humans—and specifically workers—as biological entities as well. In his view, industrial accidents were not just problems of labor, class, or management. McEvoy emphasized the parallel between bodily hazards and environmental pollution. He advocated taking an "ecological approach to industrial health and safety."[55] Put another way, workers and their bodies are as biological as trees turned into milled lumber or fish transformed into frozen meals.

Over the following two decades, a number of environmental historians, including those also working in the history of medicine and of labor, developed studies that framed the human body in, and as, environmental history in diverse cultural and historical contexts—from nineteenth-century North America to industrializing Japan and the nuclear accident at Chernobyl in 1986 (discussed further in chapter 4).[56] These works have traced the combined impact of industrialization on human and nonhuman environments, thereby challenging this distinction.

At the same time, these studies have shown how the effects of industrialization on the human body are not always experienced equally across humanity.[57] In other words, the costs and benefits of industrialization have been unequally distributed over time, space, and demographic groupings. Environmental historians, anthropologists, geographers, sociologists, and activists have teased out how class, race, ethnicity, gender, and global divides, among others, can shape access to environmental goods such as clean water and healthy food, and exposure to environmental hazards such as carcinogens and lead.[58] Environmental historians have taken up the politics of national parks and other protected areas, showing how white, middle- and upper-class Americans benefited from these landscapes of leisure, especially as escapes from the hazards of urban, industrial life. Yet wilderness protection usually came at the expense of dispossessed Native Americans and local working-class people who had historically depended on those landscapes for their livelihoods.[59]

At times, the environmental externalities of industrialization fell disproportionately on certain groups of people. For instance, the development of a radiation exposure standard based on "as low as reasonably achievable" during the height of the Cold War meant that African laborers mining uranium ore were subjected to higher doses of radiation than mine workers in the First World because, so the argument went, African countries were socially and economically unable to uphold higher norms. As a result, African miners, their families, and neighboring communities suffered the consequences of such differential standards.[60]

As these brief examples suggest, the protection of some environments and people in the industrial age has come at the expense of others. Environmental justice activists and scholars studying the history and politics of these movements have revealed how economic and political power mediates interactions with the environment to the benefit of some and the detriment of others.[61] This discussion of the *socio*-environmental dimensions of industrialization has important synergies with the chapter on the body, and we elaborate these and related themes there.

## *Limits*

As we have seen, the extensive socio-environmental dimensions of industrialization—in the form of energy and resource inputs, and the by-products of industrial processes harmful to humans and nonhumans alike—suggest how mass destruction is inherent in mass production and mass consumption, central premises of the second Industrial Revolution.[62] By focusing only on production and consumption, it is easier to perceive industrialization as modern progress and as the result of remarkable technical innovations by brilliant inventors. Adding mass destruction to this story complicates, even undermines, such an interpretation. It reveals how large-scale environmental degradation and human health effects are, in fact, intrinsic to industrial modernity. Considering production, consumption, and destruction together therefore alters our understanding and interpretation of industrialization.

At the same time, emphasizing industrialization's ecological "impacts" and destructive tendencies implies that humans alone have historical agency. The nonhuman world is simply a passive backdrop to people and their actions. Some envirotech scholars have pushed back against such claims, instead emphasizing how the relationship between humans and the environment is more reciprocal.

As we've begun to discuss in this chapter, environmental factors and processes constrained and thus shaped industrial development and use, thereby challenging the idea that nature is merely a passive stage for the human motor of history, including the history of industrialization. Indeed, ecological dynamics help explain why many key features of the Industrial Revolution occurred in Europe, and specifically in Britain. An energy crisis throughout eighteenth-century Europe was caused by deforestation, growing wood scarcity, and declining stores of coal ore in surface mines.[63] World historian Kenneth Pomeranz has argued that the relative abundance and accessibility of coal in Britain made a key difference in the industrial potential of that country versus China.[64] Certainly, environmental explanations of the Industrial Revolution can go too far. Overemphasizing ecological contexts and contingencies obscures how certain people pushed for and benefited from industrialization.[65] Forest, coal, and water resources, when combined with other significant historical processes such as enclosure, capitalism, and empire, better account for why the Industrial Revolution emerged when, where, and how it did.

Turning from the environmental context of industrialization's emergence, we may also consider how ecological factors mediate the processes and dynamics of industrial production. For example, canals and irrigation systems suggest how environmental processes still shaped and ultimately constrained their industrial objectives. River valleys were wrought by floods despite dikes and dams. At times, droughts reduced hydroelectric production, no matter how modern or how big the turbines.[66]

Another insightful example comes from the work by historian of the environment and technology Edmund Russell. Russell has demonstrated how the long fibers of New World cotton, cultivated and bred by indigenous peoples over centuries, made it much easier to mechanize textile production in the eighteenth and nineteenth centuries. In fact, the industrialization of the cotton textile industry probably would have been impossible without the particular characteristics of these cotton species.[67] The human-aided evolution of cotton species in the Americas thereby helped enable industrialization. Technological innovation ultimately depended on evolutionary change; put another way, technology relied on nature.

Yet this example also serves as an important reminder that "environmental" factors—cotton species, in this case—are not wholly natural. In this case, the longer fibers of New World cotton resulted from genetic variation ("nature") *and* the selection of these traits by Native Americans ("culture") over generations. At the same time, identifying the vital role of previously unappreciated environmental aspects in the industrialization of cotton does not diminish the importance of other social, political, and economic factors, such as European colonization of the Americas, the Atlantic slave trade, or plantation agriculture centered on the institution of slavery. The case of industrial cotton nicely illustrates not only a central premise of environmental history and science studies—namely, the concept of nature-culture—but also envirotech scholarship: the entanglement of nature and technology.

Recognizing that nonhuman entities and processes influence—but do not determine—industrial development, use, and change is important for several reasons. It strengthens ongoing scholarship, largely by social and feminist historians, that has decentered the elite male inventor as the primary driver of technological development.[68] The idea of environmental shaping of industrial technology and production expands our common definition of historical actors to include nonhumans or, in STS scholar Bruno Latour's term, "actants."[69]

This argument also dovetails with historian of technology Thomas Parke Hughes's influential concept of technological systems, but sees the environment as integral, rather than external, to those systems. In this way, scholars working in envirotech further expand our understanding of system to include natural, as well as social and technological, factors.[70] In general, thinking about industrialization in terms of complex social, technological, and environmental dynamics reflects broader discussions in the humanities and interpretive social sciences about agency that have moved toward more complicated, distributed, and reciprocal understandings of how change occurs in the world.[71]

## Temporality and Geography

We also argue that industrialization both reflected and depended on significant transformations in human-natural dynamics temporally and spatially. Considerable research has explored changes in social relations both required and produced by industrialization—most famously class and labor, as articulated by Marx and Engels.[72] But industrialization also depended on and enacted critical shifts in human-natural dynamics. Two key aspects to these dynamics are time and space.

Environmental and world historian Robert Marks has argued that industrialization marked the end of what he calls the biological old regime and brought about a temporal rupture in human-natural relationships. In *The Origins of the Modern World*, Marks argues that the biological old regime was defined by a dependency on annual solar flows, and that preindustrial societies were directly connected to and therefore depended on the environment. Some energy could be stored in both short and longer terms by reserving seed for future sowing, weaving textiles and making clothes from the year's harvest, and building shelters out of wood that would last for years. Trade also enabled some communities to exceed local energy constraints. But annual solar flows ultimately presented limits for preindustrial human societies. Using current terminology, we might say that people before roughly 1750 in Western Europe and North America relied on renewable energy and had more-sustainable relationships with the natural world, because they were limited by and therefore had to adapt to annual supplies of energy.

Industrialization fundamentally altered the temporality of human-natural dynamics in the deep past. By tapping the vast stores of energy contained within fossil fuels, industrializing societies were able to end their historic reliance on annual energy flows. Instead, they could harness the energy gradually built up over vast spans of geologic time. Moreover, they consumed and capitalized on that energy in a fraction of the time it took for coal or petroleum to come into being. Industrialization therefore relied on a tension and ultimately a rupture between natural and industrial cycles of time. In other words, industrial production fundamentally conflicted with biological and even geologic reproduction.[73]

Industrialization also transformed the temporal dimensions of human-natural interactions in the distant future. Radioactive elements are the basis of the atomic age. However, many of these elements have extremely long half-lives. The development of atomic weapons and nuclear power resulted in the production of huge volumes of waste that is radioactive and will remain so for periods far exceeding the life spans of entire civilizations in human history. Substances with these properties over these vast time scales create enormous challenges, then, for the scientists, regulators, and others charged with managing and containing them.

Government officials and the scientific community have struggled to figure out ways to "teach" generations in the remote future how to avoid radioactive waste, because they assume languages will have changed by then.[74] This example points to the importance of materials and their temporality, how new industrial materials altered the temporality and effectiveness of biological processes like decomposition, and how, as a result, human relationships with the environment had new temporal scales.

Industrial technologies also altered the spatial dimensions of human-natural dynamics in several key ways. First, industrial modes of transportation—the railroad and steamship, and later the automobile and airplane—facilitated access to more-distant places.[75] Moreover, they accelerated contact, conquest, and exchange, often under highly unequal relationships.

We should note, however, that during industrialization, long-distance connections, whether elective or coercive, were not entirely new. Extensive trade networks traversed Eurasia and Africa during the early modern era. Nonetheless, industrial technologies intensified the scale and speed of such interconnections. These technologies also made it easier and faster to extract and remove resources over greater distances. Diverse resources of empire proved vital to European industrial development. Sugar from the Caribbean and wheat from New England helped feed workers in British mills, while guano (seabird feces) from the Pacific offered valuable nitrates for agricultural fertilizer and for military explosives.[76] The natures of the colonies were therefore important to the existence and operation of metropolitan economies and technologies, thereby challenging both the empire-metropole and the technology-environment binary. Industrialization—and its entanglement with emergent capitalist economies—was certainly an economic, social, political, and technological story, but it was also an environmental one. Moreover, as many of these examples show, the boundaries between these processes are unclear.

Seeing the environment as both shaping and shaped by industrialization challenges tidy categories and binaries such as nature and culture, or nature and technology. It thus reflects growing literature in a range of fields, including environmental history and the history of technology, on hybridity.[77] As a result, seeing the "technological" and the "environmental" in industrialization as fundamentally entangled pushes us, as scholars and as citizens, to rethink simple stories often told about the Industrial Revolution. This alternative thinking instead fosters greater humility, and ultimately encourages us to see humanity as part of, and dependent upon, the so-called natural world—even as we simultaneously use, manage, and transform it. Overall, an envirotechnical approach to industrialization offers a useful language that emphasizes reciprocal dynamics and goes beyond simplistic representations of causality, agency, change, or impact. Many of these themes resonate with the following two chapters, on discards and disasters.

# Chapter Three

# Discards

**W**hy do we throw away what we throw away? The act of disposal is performed daily by billions of humans trained in the manner of disposal appropriate in their cultural and technological contexts. What we dispose of and how we dispose of it are shaped by the complex systems humanity has developed to classify and manage wastes. These actions complicate the assumption that what is natural is good, healthy, or desired, and expose the great complexity of perceptions we have about what is safe or hazardous in the human-built world.

Historical approaches to discard studies include analysis of the creation of industrial wastes, the development of infrastructure such as landfills and wastewater treatment facilities, and the emergence of formal and informal recycling systems. Efforts to define and manage waste have involved the emergence of complex technological systems and eventual dependence on those systems, have raised questions of material value, and have sparked debates over measurement and assessment of waste, and even over basic definitions as to whether waste is a hazard to be avoided or an inefficient act squandering value. With each of these issues, patterns of change and continuity provide perspective on how humans determine and manage waste.

## Definitions and Theory

The words used to describe discards—*waste, trash, dirt, garbage, rubbish*—have contested meanings that reveal social and cultural attitudes toward the material we discard. *Waste* has dual meanings; it may be used to describe worthless or filthy matter, or the inefficient squandering of potentially valuable matter.

Much of the theory in discard studies emerged from anthropology, economics, and sociology. Dirt, as the anthropologist Mary Douglas wrote in *Purity and Danger* (1966), is an action. Human society's tendency to classify material into piles of worth and trash extends throughout history. Anthropologists and sociologists have used dirt as a lens through which to view society. Douglas identified the idea of dirt as part of human society's penchant for classifying materials and behaviors as acceptable or taboo. Dirt is "matter that is out of place," creating disorder and threatening the social order. Thus, dirt is a lens for viewing cultural attitudes. Douglas argued that dirt and other waste matter derive their power not simply through being waste or having a kind of negative value. Rather, as matter out of place, things deemed dirty, spoiled, or noxious carry polluting effects for the person, transmitted through bodily substance or through touching. By avoiding impure or defiled forms, practitioners shore up purity and associate themselves with holiness or sacredness.[1]

Sociologist Michael Thompson, in *Rubbish Theory* (1979), urged the study of unwanted objects, or "rubbish," in order to understand how they are defined and altered with regard to value. Understanding how societies value their material goods is important for social scientists because all cultures distinguish between the valued and the unvalued. Further, a certain degree of social consensus regarding these values is key to maintaining social order.[2]

Rubbish theory proposes three types of objects: transients, durables, and rubbish. Transients are socially visible objects with agreed-upon values at particular times. (Think of a Kelly Blue Book valuation of a used automobile, with older cars usually appraised at lower values than newer versions of the same model.) Transients decrease in value over time and have finite life spans—their value will eventually reach zero. Durables, in contrast, also have agreed-upon worth, but continually increase in value and have an infinite life span. The values of these objects are typically determined by the powerful within society. Rubbish, by comparison, is socially invisible and deemed unvalued. Thompson argued that the rubbish category is a medium for the potential rediscovery of a past transient object and its subsequent reappearance as a durable. This notion not only explains how items of declining value can eventually become objects of great worth but also illustrates a means of bypassing the control of the powerful in society determining what is valued or not.

Because Thompson's model is necessarily temporal, it offers perspective on why historians are especially important to discard studies. The history of automobility provides an excellent example. Typically, automobiles immediately lose some value as soon as they are sold (and are no longer "new"), and go on to decline in value as their parts wear out and styles change. Often they are discarded as worthless after a few years. Some, however, see a gain in value after several years or decades and become coveted vintage cars fit for refurbishing. With such goods, cultural attachments to particular makes and models of cars help determine which are "junk" and which gain in value. Similar dynamics exist in mass-produced goods that age into antique status, such as furniture.[3]

Beyond the temporal dimensions of discarded objects, historians' attention to social structure is crucial to discard studies. The classifications that Douglas and Thompson describe have informed study of the labor used to manage wastes throughout human history. In India, Hindu-inflected caste hierarchy associated the handling of garbage and waste forms with impurity and relegated this work to the "untouchables," or *dalit*. The caste system reinforced status differences for those who had the greatest ability to spiritually elevate themselves and avoid forms of contagion.

## *Materiality of Waste*

Primary discards before industrialization included biological waste, foodstuffs, and ashes from fires. Industrialization brought new scales of production and consumption, spread the affluence necessary to purchase goods, increased the number of goods designed to be used briefly and then discarded, and encouraged innovation of new materials and combinations of materials that have reshaped the waste streams of human societies. Public systems to manage the excrement of increasingly dense urban populations grew more sophisticated. Private systems for burning rubbish and hauling ashes, dumping wastes in waterways, and trading discarded materials to producing industries seeking affordable raw materials emerged in all industrial societies.

One important question is how experts defined and classified waste and its consequences. Theories of disease transmission are central to histories of science, technology, and medicine. The managerial and technological systems that developed to control diseases are vital to this inquiry. As epidemic diseases grew more severe in the nineteenth century, new cohorts of public health experts with backgrounds in medicine and civil engineering began to change local and national health policy. Debates over disease transmission raged in the late nineteenth century as proponents of the miasma theory of contagion developed sewer systems and drained stagnant water to remove noxious odors. Proponents of germ theory advocated for disinfection of water. Similar debates over managing air pollution led to greater reliance on trained engineers to develop public health standards and industrial practices.[4]

The notion of "secondary" materials reveals how terminology evolved to take advantage of value in discarded materials. In the nineteenth century, the increased use of metals meant that goods and buildings containing metals that were obsolete could be scavenged and salvaged, leading to recycling. In the twentieth century, the vast expansion of industrial production and innovation broadened the waste stream. Of particular relevance in the rise of a mass consumer economy was the advent after World War II of packaging materials made of paper stock, metal, glass, and plastic. Packaging of commodities ranging from food to furniture became more sophisticated, and hybrid combinations of wood pulp, metal, polymer, and gas produced packaging materials customized for particular functions. Cola and other soft drinks, for example, came to be packaged in aluminum cans with plastic coatings. Plastics and aluminum were also blended to produce hybrid materials for condiment packets that were both durable and flexible. Plastic sealants on paper stock allowed waterproof priority-shipment envelopes for transporting critical documents thousands of miles with little concern that rain would damage their contents. Most of these packaging innovations were designed to be used once and be discarded; hence enormous volumes of chemically complex materials accumulated in municipal waste streams.

Innovation in the twentieth century created new threats to human and ecological health. Newly fashioned materials and chemicals included chlorofluorocarbons (CFCs) used in aerosols and air conditioning. When it was found that these gasses deplete the ozone layer after their emission into the atmosphere, they were eventually banned. Polychlorinated biphenyls (PCBs) were developed to reduce flammability in electrical equipment. Over time, they were observed as organic pollutants that persisted in land and water and thus could cause cancer in animals and humans. This pollution began in the 1930s; the United States banned production of PCBs in 1979, and an international environmental treaty banned the worldwide production and trade of PCBs starting in 2004. However, the substances continue to pose hazards in land and water more than thirty years after their production was banned. The increase in the amount of radioactive materials generated since the start of the atomic age endanger the African workers who handle uranium and the indigent Japanese workers who have been mitigating the waste at the Fukushima nuclear plant.[5]

The heavy metals, rare earth metals, and flame retardants used in electronic equipment pose hazards to the air, land, water, and people who disassemble obsolete electronics. International efforts to address these problems focus on Bangladesh, Ghana, Nigeria, and China, where exported electronic waste is separated and processed, often by hand. International markets for handling electronic waste and discarded textiles, metals, and plastics often obscure the hazards of waste creation and handling by placing waste management processes seemingly out of sight and out of mind.[6] In our search for the perfect "sink" (disposal site) for wastes, the water, the air, and the land have served as our main options. Every human attempt to find a sink without consequence has ultimately failed in one way or another, but the search for the best possible sink has informed the complex ways in which we have managed wastes throughout history.[7]

## Sinks for Wastes
### Water

The history of waste management, from classical civilizations to the global waste trade of the twenty-first century, has involved repeated attempts to remove unwanted, presumably hazardous materials from society without consequence. Historians of pre-industrial societies have focused on waste practices and infrastructures having to do with human excrement. Wastewater policy and infrastructure in European societies between the fourteenth and sixteenth centuries reveals the complexities of social systems, the cultural practices for handling human wastes, and the technologies that cities invested in and developed to carry out those cultural practices. The histories of English and Scandinavian cities reveal connections between tax policies, the development of cobblestone streets and latrines, and the growth of river-cleansing and dung cart operations.[8]

Using waterways as waste sinks had unintended consequences for human health and economic activity. London's Fleet River concentrated the city's industries and population, and its use as the primary sink for excrement and garbage in the fourteenth century rendered it unnavigable. During the rebuilding of London after the fire of 1666, architects devised a system of garbage disposal at each street corner that reduced the burden on the river as a sink for garbage, making it navigable, though still dense with sewage. The industrialization of water sanitation expanded to other growing cities in Europe and the United States during the nineteenth century.[9]

The emergence of these sophisticated technological systems to manage wastes stemmed in part from a theory of disease transmission that depended on sensory evidence. English sanitarian Edwin Chadwick, in his 1842 report on urban sanitary conditions in England, advocated sewers as a way to improve the health of the poor. This recommendation reflected the then-current *miasma* ("bad air") theory of disease, according to which poisonous vapors from putrefying materials caused various illnesses. Chadwick envisioned the city as a body irrigated by water circulating through and purifying it. His model had wide influence across Europe and the Western world.[10] Technological systems, and the people trained to develop and run them, grew in prominence as fears of contagion increased.[11]

Sewers—whether separate systems for storm water and sewage or combined systems (simpler, but prone to contaminating waterways following storms)—revolutionized modern sanitation and allowed for the installation of flush toilets in residential and commercial buildings without undue odor or contagion. This advance sufficiently transformed modern life that several studies have focused on the development and proliferation of the technology. The flush toilet greatly facilitated the funneling of human excreta into sewers.

By the middle of the nineteenth century, diseases like cholera and typhoid—carried by feces-laden wastewater to wells, spigots, and waterways—gave rise to a demand for the construction of sewers that would carry sewage not only out of and away from the home, but also away from the city. These technological systems enabled societies to flush wastes even further out of sight and out of mind.[12] Indoor plumbing allowed residents to drink tap water with more confidence, and use water for cleaning clothes, washing dishes, and scrubbing rooms. These innovations raised personal and household standards of hygiene, as well as expectations for maintaining those high standards. The result was a society of newly suburbanized Americans with simultaneously higher expectations for environmental quality and higher consumption rates of water, energy, and other resources, consumption at levels that threatened their local environments.[13]

Laborers in Jerusalem (Palestine) digging sewage canals. By the time this photograph was taken in 1936, sewer building had become a crucial part of the urban public health reforms that allowed cities to grow in population and density.
*Library of Congress, G. Eric and Edith Matson Photograph Collection (LCCN 2019707584)*

Waterways remain the most heavily used channel for removing human excrement in the twenty-first century. In the late nineteenth century, municipalities also used them to dispose of household and industrial wastes, though not without problems. Ocean dumping risked the tides washing wastes back onto land. Disposal of petroleum (either by intention or through leakage or disaster) has had lethal consequences for many forms of aquatic life and can result in visible pollution, such as the fires from oil slicks on Ohio's Cuyahoga River (USA). Responses to water pollution included legislation to restrict the dumping of industrial wastes, and new technologies for disinfect sewage.[14]

Recent history reveals that designers and engineers have realized that the boundaries between technology and nature are blurred. As the history of sustainable architecture shows, envirotechnical approaches to the built environment relate to energy, including solar energy collection and passive design standards that employ better insulation to reduce a building's energy requirements.[15] Other designs seek to reduce the burdens that sewage treatment place on the environment, including living machines that employ plants to extract nutrients from human excrement before it is released into bodies of water. Such systems are not presently developed to scale; most industrial wastewater treatment involves killing bacteria with chlorine or irradiation. The resulting wastewater is a product of envirotechnical systems, a "hybrid nature" featuring complex mixes of biological and synthetic materials. A variety of unwanted chemicals and hormones pass through the water treatment systems unaffected, with potential consequences for aquatic life and human health, including alteration of reproductive systems in frogs, fish, and humans.[16]

## Air

Histories of air as a sink for waste range from concerns about local smoke and smog to current debates over global climate change caused by greenhouse gas emissions. Using waterways as sinks for discards had limits beyond sewage. Bones, metals, and other solids often blocked rivers and streams; ocean dumping did not involve blockages, but tides regularly deposited discards on shore. In search of more effective sinks, human societies turned to the air to remove unwanted matter. They did so through burning wastes. The first waste incinerator was constructed in England in 1874, and the first incinerator in the United States was built in 1885. The fast growth of incineration in the United States was halted by a growing environmental movement, which led to both legislation and powerful grassroots efforts to keep incinerators from being sited in local communities because of concern about emissions.

Both the Resource Conservation and Recovery Act of 1976 and the Clean Air Act of 1990 set strict standards for incinerators. Modern incinerators are equipped with various air pollution control devices that treat and minimize the harmful emissions from burning waste. Policy and practice relate strongly to available resources; within much of the United States, abundant land means landfills are relatively affordable to manage, and this convenience combines with concerns over air pollution to make incineration rare. Conversely, in Japan, large cities lack usable land, and incineration both saves land and supplements nuclear and fossil fuel sources of energy. Cultural and geographical contexts thus shape infrastructure.

Waste management practices can store, disinfect, or utterly transform materials. Combustion of wastes produces ash, as well as some waste products if air pollution control technologies are use. Two types of ash result from incineration: bottom ash, the residue left over from burned waste; and fly ash, which is removed from the flue gas. While bottom ash makes up 90 percent of the total ash produced, the fly ash contains most of the toxicity from incinerator waste, including heavy metals and carcinogens. Incineration entails health concerns. Flue gases contain standard air pollutants, such as particulate matter and nitrogen oxides, as well as carcinogens such as dioxin.

For most of the nineteenth and twentieth centuries, the consequences of using air as a sink fell on municipal populations and governments, which in turn grew concerned about visible air quality and its implications for human health. The use of coal for heating and generating electricity produced thick clouds of smoke. By the 1970s, tailpipe emissions from petroleum-powered automobiles made smog more evident. Cities passed ordinances to address local air pollution, including the one St. Louis, Missouri, passed in 1940 to elevate the quality of coal used in home heating. The perceived success of this ordinance led to widespread emulation by other industrialized cities around the world over the following twenty years.[17]

Global concerns over using the air as a sink became more

widespread by the end of the twentieth century, leading to an international ban on CFCs and repeated attempts to control emissions of greenhouse gases such as carbon dioxide and methane, associated with global climate change. The difficulty of containing atmospheric emissions within national borders shaped policy and technological uses; prior to global agreements on CFCs and greenhouse gasses, European nations negotiated agreements to reduce acid rain.[18]

Even the space race has had implications for discards. After more than a half century of space exploration by the United States and the Soviet Union, debris (including highly valuable rare earth metals) orbits the Earth, imperils satellites, and sometimes falls back to Earth. Ultimately, what humans create, consume, and discard is never out of sight or out of mind—even when it leaves the atmosphere.[19]

## Land

Back on Earth, land has had a long and varied history as a disposal destination. It has been used as a sink for wastes in all human societies at one scale or another. The discards of ancient peoples make up much of the archaeological record we have today. Open-pit dumping—placing wastes in a hole or on the surface of unwanted land—involves relatively little infrastructure or maintenance. However, as cities grew in density and size, open-pit dumping became less tolerable, as rats, mice, flies, and mosquitos combined with noxious odors to create nuisances.[20]

In the 1920s and 1930s, engineers in the United States and Great Britain developed the sanitary landfill, and France developed the *dépôt contrôlé* (controlled deposit). These advances were intended to avoid or control the unhygienic aspects of open-pit dumping. Sanitary landfills differ from the open pit in that dumping is performed in compacted layers, usually with a plastic barrier between the landfill and groundwater, and methods of transporting wastes to the landfill generally involve specialized trucks to reduce the chances of losing material on the way to the landfill. Workers at sanitary landfills regularly cover the wastes with soil, ashes, or dirt in order to reduce vermin, bad smells, fires, and wind-blown litter. In addition, some forms of waste compaction are practiced, and scavengers have increasingly been banned from the landfill territory.[21]

If the technology of digging a hole and dumping wastes in it may seem simple, it is complicated by the need to develop large technological systems so the hole can be sited away from communities and the wastes can be transported quickly and efficiently to the site. The first sanitary landfill in the United States opened about three miles away from Fresno, California, in 1937. Sanitary landfills were soon developed in New York City and San Francisco. During World War II, the method was practiced and improved by the US Army. Members of the US Corps of Engineers, as well as the US Public Health Service, spread this knowledge throughout the country. Afterward, and in both the United States and the United Kingdom, sanitary engineers praised the sanitary landfill as the universal solution to the waste problem. Public and private systems in the United States and Europe have evolved since World War II to include these landfills, as well as more specialized trucks, compactors, and incinerators.[22]

In recent years, awareness that decomposition of organic matter in landfills releases methane has led to concern about powerful greenhouse gas emissions, as well as to efforts to harness this energy source and direct it to homes. Landfill gas projects in the twenty-first century include those in Brazil, China, India, and the European Union. In this way, landfills join incineration as waste management techniques that can actually extract value from discards in the form of energy to power society.[23]

The record of these discards is, like the remains of prehistoric societies, available for anthropologists to study. A pioneer in this study was anthropologist William L. Rathje, who developed the Garbage Project with his anthropology class at the University of Arizona (USA) in 1971. Students undertook independent projects to explore links between various artifacts and various human behaviors, which included a comparison of garbage samples from different households, an approach that seemed to hold great promise. By 1973, the Garbage Project sorted the garbage gathered from randomly selected households into 150 specific, coded categories reflecting when and where it was collected, its weight, and any information on its materiality available from package labeling. In 1987, the Garbage Project progressed to excavating Tucson's landfill, taking core samples of the city's garbage in order to assess changes over time. Aside from documenting the community's growth as evident from the expanded volume of garbage, the project noted the escalation of single-use packaging, especially plastic, in the waste stream.[24]

## Waste Management as Economic Opportunity

Whether society chooses to "throw away" discards into a sink or recycle them in new economic functions, the technological systems we have developed require considerable human labor to achieve these goals. The stigma associated with handling wastes has exposed societies' most vulnerable people to the environmental and health burdens of this work. The complex waste removal structures have, however, offered business opportunities for those willing to do work others avoid. In the nineteenth century, night soil men made their living by removing excrement from Chinese cities and carting them to agricultural fields in the city's hinterlands. Workers in Parisian sewers performed tasks most French people found disgusting, yet essential to improving their way of life.[25]

Much of this chapter's discussion of discards involves waste defined as a hazard to be removed. A second definition of waste relates to inefficiency and the squandering of value. Reflecting this definition, industrial society identified discarded materials that could be used as inputs in production. By collecting and reusing these discards, industrial firms could increase profit.

Opportunities for immigrants to start businesses in waste handling, sorting, processing, and selling in the United States existed between the Civil War and World War II, because native-born Americans who had other options preferred not to do such dirty work. Some of these were small, private businesses handling ashes or trading scrap materials. Families of first- or second-generation immigrants often ran these businesses, and

Men, women, and children pick through a junk pile in New York City in this photograph by Frank M. Ingalls (undated, but from his collection from circa 1901–1930). In this way, "pickers" suggest how "junk" or waste could actually have value, thereby challenging our initial understanding of "waste." *Collection of the New-York Historical Society (nyhs_PR028_b-02_f-08_305-01)*

successful operations were passed down through the generations. Others were cooperative arrangements between waste handlers that provided stable employment to generations of workers within the participating families.[26]

Because of the durability, malleability, and construction suitability of metals, their recovery and recycling have remained ubiquitous practices into the twenty-first century. Earlier, when recycling laws focused more specifically on hygiene, cities recycled materials besides wood and metal. As the industrial era began, metal, textile, and paper supplies grew in distribution. Spain built copper-recycling facilities in response to continental appetite. England developed rag recycling in response to aristocratic regulations on textiles. So-called dust yards sprang up in a number of European cities where debris was gathered from city streets, sorted by human hands, and sent to second-hand merchants. The residual dust and cinders from coal and wood fires were sent to fertilizer makers. The quest to wrest value from discards produced formal and informal networks of labor and technology for salvage and transformation. Eventually, increased industrial-era waste could be attributed to conspicuous consumption of ever more goods and products.

Ubiquitous waste precipitated municipal health laws crafted to protect citizenry from the by-products of disposal. These laws also helped improve the health of dust yard sorters and made citizens generally aware of disposal practices, including that of reusing much of what had been previously discarded. As economies industrialized, they encouraged formation of small businesses devoted to reclaiming rags, metals, and other goods from consumers and returning them to paper mills, steel mills, and other manufacturers willing to pay for secondary materials.

After World War II, waste management became more corporate, and companies such as Waste Management and Browning Ferring Industries made billions of dollars handling the discards of millions of people. One scrap-recycling firm, Luria Brothers, became the dominant broker of scrap iron in the United

This photograph taken by John Collier, Jr., for the US Office of War Information in 1942 shows residents of Fort Kent, Maine (USA), collecting scrap metal that will be melted down and used in military production. World War II scrap drives were high-visibility examples of large-scale salvage of discarded materials. *Library of Congress, Prints and Photographs Division (LCCN 2017823621)*

States after World War II and was ultimately sanctioned by the Federal Trade Commission for uncompetitive business practices. Waste Management grew from one Dutch immigrant's Chicago-based ash-hauling business in the 1930s into the largest waste-handling firm on Earth, disposing of garbage and yard waste and recycling for thousands of municipalities around the world by the end of the twentieth century.

The capital available to such giants allowed research and development into new technologies. Important twentieth-century technological developments included sanitary landfills that superseded open-pit dumps and sought to protect groundwater from disposed garbage, large incinerators that reduced the physical volume of garbage, and giant shredders that separated out the recyclable steel in automobile bodies. While technological advances could provide a competitive advantage in increasing the speed and volume of handling, they could also produce unintended consequences. The automobile shredder was credited for returning as much as one-third of the United States' recycled steel in 1980; its operations produced hazardous wastes that led to a series of fines and yard closings in the 1990s.[27]

Municipalities also created public departments to manage sanitation. Col. George E. Waring was an important figure in this development in the late nineteenth century. The epidemics that threatened most cities in the 1800s were especially devastating to Memphis, Tennessee (USA), which was struck by yellow fever epidemics in the 1870s. In 1878, more than one-sixth of the city's population died of the disease. The following year, civic leaders repealed the city charter and established a commission to govern the city and rebuild its sanitary systems from scratch. Local and state officials invited the National Board of Health to investigate and make recommendations. Waring proposed that Memphis build—with public funds—a then-unique sewer system to regularly discharge household sewage.

Waring's success with organic wastes led to his 1895 appointment as street-cleaning commissioner of New York City, where he was tasked with removing organic and inorganic wastes from the dense city streets. Waring directed households to use a "primary separation" system in which garbage, rubbish, and ashes were kept in separate receptacles awaiting collection. The

Street Cleaning Department could then easily use different methods of disposal for the different waste materials.[28]

Waring used a variety of waste disposal methods that used the land, air, and water as sinks, and employed machinery and humans in an evolving, responsive system of removal and reuse. He continued the practice of ocean dumping; he also commissioned new types of scows to dump wastes farther from shore and supposedly reduce the chances of their washing up on beaches—measures that were only partially successful. Under Waring, New York City built the first municipal rubbish-sorting plant in the United States, where immigrants handpicked salvageable materials out of the discarded rubbish for resale. Profits offset the city's collection costs.

Waring's modernization of New York City's Department of Streets and Sanitation expanded under his successors to encompass more than ten thousand workers in the early twenty-first century. Anthropologist Robin Nagle notes that the tasks carried out by this workforce included not only disposing of household waste but also salting roads and removing snow after blizzards. The workers' expertise managing the urban environment provides specialized value during extreme weather events. Nagle's description of the ways workers' expertise develops offers further opportunities to historicize waste workers' experiences as mediators of nature and technology.[29]

Less formal systems of waste management also employ thousands of workers. In megacities around the world, marginalized people subsist on income from waste handling. Some make homes in the landfills they scavenge, creating not only economic value from commodities they find but also subsistence shelter.[30]

The embodied aspects of waste invite historians to analyze the social and cultural structures that Mary Douglas argued are embedded in all human societies. Embodied environmental history (discussed further in chapter 5) analyzes the consequences of exposure to the materials classified as dirty and dangerous enough to warrant removal. Within the United States, attitudes toward hygiene became central to white identity after the abolition of slavery. These attitudes in turn this racialized waste management occupations and reorganized the geographies of waste within the country between the Civil War and 1970. The systemic hazards that waste pose to the body inspire resistance movements against, for example, spatial inequalities of pollution exposure in residential neighborhoods of industrial cities, and the proximity of waste management facilities to majority–African American residential neighborhoods in the United States. During the 1980s the environmental justice movement was started by individuals, primarily people of color, who sought to address the inequity of environmental protection in their communities. The work of handling waste became racialized in the United States after the Civil War; such systemic inequality raises questions about how the formal and informal systems of managing waste organize technology and labor in ways that have unintended consequences that harm vulnerable people.[31]

Scavenging involves identifying value in the discards of others, recalling Thompson's observation that value is socially constructed. The stigma that many scavengers around the world face recalls Douglas's attention to waste classification and fears of impurity. In working to extract value from discards and return

This Gordon Parks photograph from 1942, entitled "American Gothic" to evoke the more famous Grant Wood painting from 1930, depicts Ella Watson. At the time of this photo, she had worked as a "charwoman" for the US government for twenty-six years, cleaning offices, hallways, and bathrooms in office buildings. Human labor is a crucial component of sanitation systems, and the history of sanitation in the United States involves labor regimes that place disproportionate burdens on women and people of color. *Library of Congress, Prints and Photographs Division (LCCN 2017765074)*

them to industrial production, scavengers are vital to the industrial recycling of metals, textiles, and rare earth minerals.

Formal, industrial-scale recycling can recover substantial amounts of material, reducing the need to extract new raw materials such as bauxite for aluminum and rare earth for electronics. In some industrial sectors, such as automobile manufacture, standardized mass production can yield economies of scale through harvesting components from junked products. But industrial recycling generally, and that of electronics specifically, involves energy- and material-intensive processes. It often requires transportation over substantial distances from collection to processing facilities, which adds to its overall environmental impact. The operation of smelters involved in recovering material and energy from electronics can release toxic substances, such as lead.

These systems work best when the collected materials are easily, quickly, and affordably separated and sold for reuse in industrial production. Collection of metals, particularly steel and aluminum, has historically been the most effective of these programs; salvage of more complex materials such as plastics and integrated electronics has proved more problematic. There are limits to the environmental benefits these programs generate. Significantly, they have shifted the burden of responsibility for removing waste stream materials from the producers who design and manufacture products to the consumers who use them.[32]

All of these efforts to reclaim waste require the development and maintenance of complex technological systems. Municipal recycling programs designate specialized trucks and work crews to collect curbside recycling. The trucks transport the materials to material reclamation facilities (MRFs) that may use conveyor belts to facilitate separation of different materials. Private and public scrap recycling also involves shredders, balers, magnets, and shears to separate and shape materials into forms easily transported to and used in industrial production facilities. Aside from the trucks collecting household discards on city streets, consumers may encounter other technological innovations to assist material recovery. The reverse vending machine, used initially in Scandinavian nations in the early 1970s, and thereafter in parts of the United States, provided consumers who brought their beverage containers to the machine with immediate economic compensation. This technology was most effective when joined with "bottle bills," or deposit laws establishing fixed values for cans and bottles that consumers would get back when returning the containers. Beverage industries opposed such legislation, largely stopping the spread of such laws beyond the thirteen states that had enacted them by 1987.[33]

The early twenty-first century saw attempts to shift responsibility for manufactured products at the ends of their useful lives onto producers. In the United States, the 1980 Comprehensive Environmental Response, Compensation, and Liability Act (CERCLA), which created the so-called Superfund, made potentially responsible parties liable for the mitigation or remediation of hazardous waste sites. Other solutions falling onto the designers and producers of products include design for disassembly, "cradle to cradle" certification, "extended producer responsibility," industrial attempts to upcycle (produce goods of greater value from discards), and related "closed-loop" approaches to design, construction, and disposal. Industrial management of wastes may involve shipments and processing that transcend political borders; the technological and political complexities of electronic waste handling on a global scale involve systems spanning thousands of miles over land and water.[34]

These approaches are rooted in the idea that recycling can take discarded materials out of the waste stream. As industrial society entered an environmentally conscious era in the late 1960s, recycling became part of an environmentalist ethic, and thousands of municipalities developed curbside recycling collection systems in the final third of the decade. The opportunities and limits of waste reduction through recycling reveal the complexities of consumption, systems of industrial production and distribution, and dependence on systems of waste management that place most of the burdens of waste on individuals and governments, instead of on the industries that design and manufacture the materials ultimately discarded.

A distinguishing feature of the systems that industrial societies have designed to move wastes out of sight and out of mind is the distribution of waste's burdens onto vulnerable peoples. Within the United States, the environmental justice movement in the late twentieth century was catalyzed by protests against a hazardous waste facility's siting in a rural African American community in Warren County, North Carolina. Historians who have probed the socioeconomic dimensions of waste management have provided context for the conditions informing the movement.[35]

The advances in transportation that accelerated the Industrial Revolution also affected the history of waste management. The twentieth century saw rising exports of actual materials and practices, forming transnational systems that placed the burden of storing and processing automobile batteries, computers, and municipal solid waste beyond the borders of wealthy industrialized nations. The most prominent example at the turn of the twenty-first century involved electronic waste flows away from the United States and Northern Europe to regions in East Asia, Africa, and Latin America. Scholars have also documented waste created in the United States going to Mexico and Canada for siting, and the systems of recycling metals, plastics, and paper stock regularly involve material flows over thousands of miles across land and oceans.[36]

States and corporations have attempted to curb waste creation at the design and production stages. The European Union's "extended producer responsibility" (EPR) regulations under which manufacturers have financial or physical responsibility for the treatment or disposal of postconsumer products, have affected the design of packaging and electronics so that disposable products can be reused, and have addressed the difficulties of recycling materials from discarded goods. Industrial society's public and private means of managing wastes continue to add complexity in ways that show the value of histories of nature and technology.[37]

Human societies have increased in size and complexity over history. Humanity grew to more than seven billion people inhabiting the planet in 2012, but that is not the only reason that the effects of our consumption and disposal are more visible than ever. The innovations that have made modern life more comfortable and convenient to the people who have access to modern amenities involve the extraction of resources, the perpetual innovation of new goods and services, and the resulting disposal of a wider variety of materials and material combinations.

Our innovations have made the classification and management of waste—practices undertaken by all human societies—more difficult, especially as the scale and complexity of our disposal have grown. Some of this disposal is readily measurable, as with the emission of carbon dioxide into the atmosphere. Concerns about this emission led to the creation of 350.org in 2007, an advocacy group using the measure of 350 parts per million of carbon dioxide in the atmosphere as the appropriate target level for reducing the effects of climate change.

Other forms of disposal do not lend themselves to easy measurement or assessment of consequences. Since Capt. Charles Moore first observed vast expanses of plastic wastes floating in or sitting at the bottom of the Earth's oceans in 1997, the plastics in the world's waterways have defied easy classification or measurement. Degradation paths of different plastics vary, as do the volumes proliferating in those waterways. Max Liboiron notes that ocean-borne plastics are a "wicked problem" endangering life both in the water and on the land. The scale and complexity of this problem do not lend themselves to easy classification or solution. Waste on this scale is neither out of sight nor out of mind, and is a challenge to societies' attempts to minimize the hazards of waste. Historical focus on temporal change and continuity allows us to examine the scale and seriousness of the enduring problem of waste.[38]

# Chapter Four

# Disasters

W hat *is* a disaster? The term generally suggests a dramatic, calamitous event with a clear beginning, a discrete cause or set of causal factors, and a limited duration that captures public attention. Consider the 1755 Lisbon (Portugal) earthquake, the 1910 Paris flood, the 1911 Triangle factory fire in New York City, or the 1984 explosion at a Union Carbine chemical plant in Bhopal, India.[1] Or, more recently, September 11, 2001; Hurricanes Katrina and Rita, which hit New Orleans (USA) in 2005; and Superstorm Sandy in 2012.[2] Or, the 9.0 Tōhoku earthquake and the ensuing tsunami that slammed into the coast of northeastern Japan in March 2011 where Fukushima Daiichi, a large nuclear power facility, was located.[3]

Over the past few decades, however, a number of scholars have suggested that disasters are not as straightforward as they seem. Rather, they have argued that—despite the common phrase—there is no such thing as a natural disaster.[4] As environmental and legal historian Ted Steinberg memorably put it, natural disaster has an unnatural history.[5] Disasters are instead shaped by social relations and power dynamics in multiple, complex ways. As such, they illustrate the idea of hybridity. They are, therefore, ultimately sociopolitical problems.

During Hurricane Katrina, for instance, residents of New Orleans were unequally affected by the storm and levee failure. Historic patterns of settlement, urbanization, and race relations meant that the city's poorest residents, primarily African Americans, lived in the parts of the city at the lowest elevations—the region most vulnerable to flooding.[6] As another example, in 2011, the Japanese government declared a nuclear emergency and established an evacuation zone around Fukushima Daiichi, but it repeatedly modified the evacuation's extent and terms. Given that other nations' governments with citizens living in the country issued stricter protocols, it appeared that the Japanese state was setting optimistic regulations in order to downplay the severity of the crisis.[7]

These brief examples suggest the complexity of disasters, as well as their historical and political dimensions. Examining the politics of disaster means opening up questions and categories previously assumed clear-cut. For starters, paying close attention to politics changes our initial question. Instead of asking, What *is* a disaster? we can—and should—ask, *Who gets to define what a disaster is?* Put another way, what counts as a disaster?[8] And which disasters are, as a result, (made) invisible and therefore ignored?[9] How might reconsidering the time frame of disaster raise new questions and open up new insights? And how do we make sense of the multiple factors that often contribute to disasters? In short, disasters particularly and powerfully exemplify the central concepts of porosity, hybridity, and system.

Historians of technology, environmental historians, and those working at the nexus of these fields have played key roles in this flourishing debate. Steinberg, John McPhee, and others established critical foundations for what is now called disaster studies, but work in this area has grown considerably since the turn of the millennium, due in part to recent events. Notably, to a greater degree than before, disasters such as Fukushima Daiichi and Superstorm Sandy affected historically privileged groups. In response, more scholars in the global North have seemed to turn their attention to these issues. Indeed, some scholars in the history of science and technology and STS have called for research that speaks directly to contemporary concerns.[10] Kim Fortun and Scott Frickel have proposed "disaster STS," historian of technology Scott Knowles has pushed colleagues in that field to actively engage with and intervene in the development of disaster policy, and historian of science Naomi Oreskes, who has written extensively on the history and politics of climate change, recently defended presentism as a driver of historical inquiry. The emergent field of disaster studies is a logical outgrowth of her call.[11]

Over the past fifteen years, these scholars, among others, have developed new insights and new conceptual toolkits by examining diverse disasters. Three major themes in this recent work are temporality, envirotechnical disasters, and the politics of causality, blame, and attribution. All of these themes both build on and refine existing work at the intersection of environmental history and the history of technology. We discuss each of these themes in turn.

## Temporalities and Slow Disaster

When we think of a disaster, many of us probably call to mind events like Chernobyl, a nuclear accident in April 1986 in Ukraine, then part of the Soviet Union. Ironically, the accident at the Chernobyl Nuclear Power Plant started during a safety test. Technicians were simulating a power outage to make sure that the reactors would continue to operate safely. Instead, a failure to follow testing protocol, when combined with reactor design flaws, resulted in a sudden release of energy, vaporizing superheated cooling water and ultimately rupturing the core of the No. 4 reactor in a steam explosion. The reactor core burned uncontrollably, releasing airborne radioactive contamination for nine days across parts of the Soviet Union and Western Europe before it was finally contained.[12]

The temporal scale of disaster illustrates how common assumptions spotlight certain calamities—like Chernobyl—while rendering others imperceptible. Although disaster usually implies an acute catastrophe and tends to center on a specific, discrete event (a fire, a flood, a nuclear meltdown), historians and other scholars have suggested that we should think about disaster not only in these ways but also as a chronic, gradual, long-lasting process. To paraphrase environmental humanist Rob Nixon, we know a lot more about spectacular disasters than we do about the "slow violence" of other disasters—disasters that are in fact rarely thought of as such.[13] Knowles builds on Nixon's work to ask, What if we do not assume that disaster is an isolated, one-time event? In other words, what if we challenge the idea of "the event" itself? For example, asking whether the wind, rain, storm surge, levee breaches, the Federal Emergency Management Agency's response, or racial and class inequality was the "real" cause of crisis in New Orleans during Hurricane Katrina reproduces the assumption that a disaster is caused by one factor or singular event.[14] Indeed, the single word *Katrina* now signifies all this and much more.[15] By reconsidering the temporal frame of disaster, scholars such as Nixon have begun teasing out new insights.[16]

For one, some disasters may transpire in time frames radically different from one another: the acute crisis of Bhopal versus the chronic effects of industrialization and widespread chemical use over decades discussed in Rachel Carson's *Silent Spring*; the immediacy of the BP Deepwater Horizon blowout in the Gulf of Mexico versus the decades-long Guadalupe Dunes oil spill in California; or, perhaps most starkly, the apocalyptic drama of a Chernobyl or Fukushima versus the enormous, mind-boggling time scales of radioactive waste stored in various facilities around the world.[17] These brief comparisons suggest how gradual, incremental changes can be disastrous, even though they may lack a dramatic onset, singular event, striking visual impact, or immediate grand scale.

These examples show how small-scale changes can be so negligible and so slow that they incite little concern. Some remain undetected for years. Collectively, though, they can add up and result in what some ecologists call a diminished baseline. Or, in human health terms, these changes may bring about a quiet epidemic, like the global asthma epidemic. However, because they are gradually insidious, they usually receive less attention.[18] In other cases, an acute crisis such as the 7.0 earth-

Martin Stott photographed the community affected by the Union Carbide explosion in Bhopal, India, about one year after the disaster of 1986. The abandoned factory is in the background of this photo. Graffiti, in the foreground here, including "Carbide thrive, people die," suggests how political and social movements had by 1987 emerged in response to the disaster. This photograph suggests the complex temporalities of disaster—from the explosion and immediate illness and deaths to unemployment, ongoing protests, and latent diseases. *"Protest Graffiti on Union Carbine Rd.," photograph by Martin Stott, included in Bhopal Medical Appeal Flickr account, https://www.flickr.com/photos/bhopalmedical appeal/14283708466/in/photostream/. Creative Commons Attribution–Share Alike 2.0 Generic license. No changes have been made.*

quake in Haiti in 2010 may indeed be framed as a disaster. Yet it can be followed by the slow violence of cholera outbreaks, protracted rebuilding efforts, and the politics of international development aid—not to mention the long-lasting legacies of slavery and colonialism.[19] These aspects are as important to consider as the earthquake itself, though it may serve the interests of some people to highlight certain aspects of a disaster over others.

Similarly, climate change offers a timely illustration of how the violent crisis frequently obscures the mundane. Stark illustrations of fossil fuel dependency such as the BP Deepwater Horizon oil spill or debates over the Keystone XL Pipeline capture popular, media, and political attention. Yet more incremental historical processes such as suburbanization, which in countries like the United States is centered on the automobile, play vital roles in perpetuating fossil fuel use and therefore fueling global warming. In other words, the Deepwater Horizon blowout may receive considerable attention—and rightfully so. Yet car-dependent suburbs, daily commutes, and the mundane chore of filling up the gas tank do not—even though, collectively, such routine, even banal habits are major contributors to climate change.[20]

Examples such as the asthma epidemic and Haiti's earthquake therefore expose not only the tragedy of slow disaster but also the structural violence of defining and perceiving only acute, dramatic crises as disasters.[21] Michelle Murphy's concept of "regimes of (im)perceptibility" helps us understand how current knowledge regimes focused on disaster are more effective at bringing some catastrophes into sharp relief than others. Murphy explores how arrangements of discourses, objects, practices, and subject positions work together within a particular discipline or knowledge tradition to shape what is perceptible, knowable, and therefore known—and, simultaneously, what is

This photograph by Carol M. Highsmith, which she captioned "View of the maze of freeways and railway lines that intersect in downtown Dallas, Texas" (2014), suggests the extent of carbon-based transportation infrastructure in places like Dallas, Texas. *Library of Congress, Prints and Photographs Division (LCCN 2014632611)*

imperceptible, unknowable, and thus unknown. In other words, producing or clarifying knowledge in one area means ignoring or obscuring knowledge in others.[22] Current knowledge regimes regarding disaster favor those with spectacular catalysts like Fukushima's 9.0 earthquake, climactic moments like the reactor explosion at Chernobyl, or vivid, shocking imagery such as vast oil slicks and oil-covered birds beached on the shores of the Gulf of Mexico after the Deepwater Horizon blowout. In other words, these regimes favor the stunning over the banal; the sudden onset over the ongoing problem; the large-scale crisis over the incremental challenge; the devastating effects of a "natural" event or the calamitous failure of a "technological" system over disasters caused by multiple messy, interrelated factors that are more difficult to pinpoint (and, as we discuss below, often more contentious). In brief, they favor the acute disaster over the chronic predicament. Accordingly, ongoing conditions, gradual changes, and mundane practices—what we might call quiet crises—are less perceptible, sometimes even invisible, in these regimes.[23]

As Nixon notes, slow disaster remains less visible, in part because of competing time frames. Corporate media thrive on featuring (and then promptly forgetting) the latest crisis *du jour*. Given the speed of the internet and social media, it may in fact be the crisis *de la minute*. Meanwhile, the temporal rhythms of political campaigns push short-term promises with the hope of Election Day returns. Then politicians face ticking clocks to make good on their promises before the next election. Yet cycles of biological, ecological, and evolutionary time do not necessarily match up with either cultural or technological change, not to mention human generational time, over which environmental costs may be deferred and born by future generations. Historical perspectives help us see and understand the tensions among these time frames.

Indeed, some socio-environmental costs may not be apparent for years. The example of endocrine disruptors (chemicals that interfere with hormonal systems) shows how these costs can be experienced not only by the current generation but also by its children and even grandchildren. Government regulations often come too late. At times, widespread adoption of such laws predates a full understanding and therefore can lead to unintended consequences, further strengthening Nixon's point about competing time scales. When policies follow exposure, they often play catch-up—with health, illness, and real lives (both human and nonhuman) at stake.[24]

## History and Memory

Paying attention to temporality yields other insights. Slow disaster is also rendered imperceptible by what marine biologist Daniel Pauly calls "shifting-baseline syndrome" and psychologist Peter Kahn terms "environmental generational amnesia."[25] Over time, small, seemingly insignificant, yet chronic changes accumulate, such that subsequent generations may assume what surrounds them is normal and natural, rather than already degraded. Accordingly, "'The problem is that people don't recognize there's a problem' because they don't know any better."[26] For instance, as the chapter on sensescapes discusses in more detail, light pollution caused by the growing use of artificial light at night has gradually worsened since the late nineteenth century. Most people in the urban global North in the early twenty-first century have never known truly dark nighttime skies.[27] Thus, they do not perceive the slow disaster in their night skies—and environments and own bodies—because their benchmark of what the night sky should look like does not take into account substantial change over previous decades. Kahn's concept of environmental generational amnesia captures this tension between human and ecological time scales. It also shows how losing a long-term perspective can be costly, both ecologi-

cally and socially. Without knowledge rooted in the past, ecological changes become normalized and naturalized.

The history and memory of disaster (or lack thereof) can
be influential shapers of the present in several ways. What is
remembered—or not—can have tangible consequences. The New
Madrid earthquakes in 1811–12 in the Mississippi Valley, unlike
those along the Pacific coast, have generally been obscured in
US history. The American Midwest is therefore less prepared
for earthquakes because emergency response, zoning, and other
policies have not been integrated into political and community
responses.[28]

Sometimes the history and memory of disaster take the form
of institutionalized knowledge—official state investigations of a
calamity or government memorials listing those who lost their
lives. Individuals' and communities' experiences with disaster,
however, may provide counternarratives that may actually critique official stories told by scientists, experts, or the state.[29] For
instance, many Japanese people filtered the experience of the
triple disaster at Fukushima Daiichi through the lens of past
crises in Japan, including the atomic bombings of the cities of
Hiroshima and Nagasaki during World War II and the mercury
poisoning at Minamata, caused by the release of methylmer-
cury in industrial wastewater from a Chisso chemical factory
into Minamata Bay.[30] For some living in 2011, these were memories from childhood or adolescence; for others, it was history
learned and retold in school, textbooks, museums, and family
lore. These kinds of histories and memories matter because
they often shape how people perceive and explain subsequent
disasters.

## Envirotechnical Disaster and the Politics of Causality

History and memory shaped the ways in which many Japanese people interpreted the triple disaster at Fukushima. Sara Pritchard has argued that we should interpret it as an *envirotechnical* disaster: a complex—and ultimately tragic—conjuncture of environmental, technological, and sociopolitical factors. Such an argument, like Hughes's systems theory and sociologist Charles Perrow's notion of normal accidents, questions the tidy categorization of causal factors and simplistic notions of linear causality.[31] In contrast, the argument is predicated on the ideas of permeability, hybridity, and systems. For many people in Japan, however, these more-complex models of causality that point to multiple, dialectical factors risk obscuring—and thereby diminishing—corporate and government responsibility. For them, hybridity is a politically dangerous argument. They believe it depoliticizes disaster and ultimately disempowers those who already have less voice and influence.[32]

As these critics suggest, some people and groups have strong incentives to distinguish multiple factors and to point fingers in certain directions and, simultaneously, not in others. In particular, they may strategically shift attention from sociopolitical causes like poverty or nationalism to "natural" elements like an earthquake or tsunami or "technical" issues like flooded generators. After all, "it suits some people to explain [disasters] that way."[33] At Fukushima Daiichi, for example, the more the Japanese government and Tokyo Electrical Power Company (the agency in charge of the facility) could blame the earthquake and tsunami for the crisis, the less they were responsible for any mistakes in designing, building, operating, or regulating the plant. It was therefore in their political and economic interest to ignore other social, political, and cultural considerations, such as the decision to build nuclear reactors on the seismologically active Pacific "Ring of Fire" (the edge of the Pacific Ocean where earthquakes and active volcanoes are common) or the importance of nuclear power to Japanese "technological nationalism" from the 1950s through the 2000s.[34] Such examples indicate how we cannot simply accept accounts of either "natural" disasters or "technological" failures without asking questions. To the contrary, as Knowles argues, "we should document the naturalization process of natural disaster and the deterministic constructions of technological disaster."[35]

As Knowles's words suggest, the stakes of apolitical accounts of disaster are huge. Knowles denounces the "societal failure to imagine disasters as political processes." We can go even further than this condemnation. Powerful groups have strong interests in portraying disasters as natural or technological precisely because these categorizations obscure political, economic, and cultural factors, thereby absolving certain people or groups of responsibility and blame. "The apolitical rendering of disaster is a necessary act," then, for those who wish "to allow the risk-taking of late capitalism to proceed apace unchallenged."[36] In this sense, we need to be skeptical of actors' claims regarding environmental or technological causes of disaster because they can be expedient and strategic. Indeed, huge profits (for some) can be made from disaster, in what Naomi Klein calls "disaster capitalism."[37] Similarly, actors emphasizing the porous boundaries between nature and technology can naturalize more-substantial

US Federal Emergency Management Agency photographer Andrea Booher took this photo of Dannie Randal, a resident of the Ninth Ward in New Orleans, on October 18, 2005. Randal was visiting his family's home for the first time since Hurricane Katrina hit in late August of that year. The photo depicts the extent of damage to homes and possessions, as well as the risks posed to human health during cleanup. The city's African American community, disproportionately located in New Orleans's lower elevations, including the Ninth Ward, were also disproportionately affected by the disaster. Hurricane Katrina is a prime example of environmental injustice. *US National Archives and Records Administration* (5693680)

interventions, such as those in France's Rhône River valley in the late twentieth century. Thus, we should be wary of enviro-technical arguments by historical actors as well.[38]

Apolitical accounts and representations of disaster are forms of "organized ignorance." Employing this concept, sociologists Scott Frickel and M. Bess Vincent argue that a lack of knowledge is not necessarily innocent. At times, it may be an unintended accident. Environmental generational amnesia, for example, may simply be the result of gradual change over time and an absence of historical perspective.[39] But in other cases, a (seeming) lack of knowledge and the ways in which knowledge making is organized for studying certain questions—and not others—actually serves vested interests. In the case of Hurricane Katrina, compartmentalizing knowledge, such as that resulting from little dialogue between the disciplines of engineering and ecology, can have serious consequences, because fundamental connections between these discrete bits of information become hidden and thereby ignored.[40] Some people—especially more marginalized groups in society—usually pay the price for such ignorance.

Exploring disasters—whether a specific crisis such as Fukushima, a concept like Nixon's slow violence that interrogates the idea of crisis, or a general predicament facing modern societies[41]—is relevant to the themes of this book for several reasons. As we've seen, disasters powerfully illustrate the murky boundaries between technology, the environment, and society. Scholarship on both historical and contemporary disasters also highlights the politics and real-world implications of calling a disaster natural, technological, or social. Such categorizations are political and strategic.

Moreover, this research speaks to the insights gained from putting environmental history and the history of technology in conversation with each other. As the concept of envirotechnical disaster suggests, people in the past (and, until recently, many scholars) believed the environment and technology could be discussed separately. Time and again, disasters have shown us not only how erroneous but also how dangerous such an assumption is.

At the same time, scholarly studies of disaster are, by necessity, interdisciplinary; they draw from the history of science, STS, anthropology, geography, and political ecology, among others. Envirotech scholarship is especially useful in this interdisciplinary conversation because it can help us think about and describe the complex dynamics between what we call technology and what we call the environment. Moreover, historical approaches enable us to understand the historical contexts and contingencies of these relationships—the way they emerged in the past, how they help explain the present, and how they might be shaped for better possible futures.

# Chapter Five

# Body

**W**hat could be more natural than the human body? Human beings, like other organisms, are indeed products of nature. Skin, eyes, hair, internal organs, and limbs exist because of biological processes. Yet our bodies are also contested arenas for the mediation of technology and the environment. Histories of the human body reveal the lenses that scholars of nature and technology have used to understand our world and ourselves. This chapter examines how scholars make sense of the body through its model as a system, its place within larger systems, its vulnerabilities to environmental hazards, its effects on environments, and its signified identities of race, class, and gender. The human body is a provocative area for investigating the complex interactions of technology and nature.[1]

## *The Body as System*

An enduring theme in the history of technology is the development of complex systems in what Thomas Parke Hughes called an evolving "human-built world."[2] Environmental historians have examined changes in ecosystems over time. Some of the most provocative envirotechnical analysis considers the human body as a system of interlocking organs, processes, and senses that both interpret the world and can be altered by phenomena from that world. Sensory approaches to the body—understanding history through smell, taste, sight, and the range of touches (as discussed further in the sensescapes chapter)—allow for a perspective on the ways by which humans have attempted to classify and relate the different senses and organs in the human body.

The main literature that seeks to understand the body as a system involves the history of medicine. The most-established methodological approaches to the body have emerged in the history of medicine and in feminist analyses of the body's role in society.[3]

A particularly contested aspect of this history involves analysis of understandings of sex and gender from antiquity to the modern era. From classical Greece through European medical communities of the nineteenth century, the female body was conceived as the inverse of the male. Medical understandings of sexual divergence developed over the twentieth century, but male doctors ascribed a variety of psychological maladies to dysfunctional female anatomy and developed technologies intended to regulate these defined "diseases" and exert social control over women.

This topic reveals power relations in constructions of medical expertise. A stark example is the machinery doctors used in the late nineteenth century to treat "hysterical" maladies, meaning physical or psychological illnesses related to suffering in the womb. The treatments of hysteria provided a gendered context for both changes in medicine and in technology related to deep unease with the role of the female body in industrial society.[4]

These historical constructions speak directly to current understandings of sexual identity and the ways in which modern society's architecture, medical classifications, and policies are built around reinforcing the binary construct of two distinct sexes, despite the presence of intersex and transgender people who do not neatly fit binary categorizations. When understandings of biology evolve to undercut design choices in the infrastructure that serves to manage biological functions, designs are likely to evolve in turn. In the United States, for example, twenty-first-century debates over access to sex-specific bathrooms for transgender people continue to illustrate the use of technologies to exert dominance over vulnerable people.[5]

The unintentional consequences of technological innovation may reshape the human body; for example, the shoes that ballerinas wear has visible effects on their skeletons. Similarly, the human body adapts to changes in furniture. One approach to the history of technology assesses such consequences, and includes studies of how different forms of seating have shaped the skeletons of humans who use the seating in different ways, and how protective technologies such as football helmets have reshaped injuries rather than eliminating them. This research indicates the unintended consequences and complexities of our intended uses of technology. Technologies transforming human bodies range from the use of brass rings to elongate the necks of Kayan Lahwi women in South Asia, to corsets that compressed the ribs and spines of women in Western societies between the

The rigors of ballet, including repetitive motion and the effects of wearing pointe shoes, not only demonstrate the physical capabilities of human bodies in choreographed productions but also transform dancer's bodies through injury. The demands of particular sports and arts produce particular risks to the bodies of participants, including knee injuries to basketball players, elbow injuries to tennis and baseball players, and brain injuries to North American football players and boxers and other participants in what are referred to as contact sports.
*US National Archives and Records Administration (7518579)*

sixteenth and nineteenth centuries. Humans seek technologies to alter the systems of our own bodies.[6]

Manipulating the body affects its form and function; in this way, such adaptations may be seen as system maintenance. Metaphors change the way we see our bodies; they also affect how we see the world we have built. The human body is sufficiently complex that philosophers and scholars of technology often use it as an analogy for built systems. Urban planners—from Patrick Geddes and Frederick Law Olmsted to practitioners of the New Urbanism (an approach valuing walkable blocks and streets with a mix of residential, retail, and public spaces) inspired by Jane Jacobs—have envisioned functioning cities as interlinking systems of transportation, energy, food, and waste management that map onto the human body's nervous, cardiovascular, and digestive systems.[7]

The body is useful for explaining concepts; as metaphor, it allows us to understand aspects of the world. Edward Tenner, for example, used the body to explain the concept of revenge effect: "If a cancer chemotherapy treatment causes baldness, that is not a revenge effect; but if it induces another, equally lethal cancer, that is a revenge effect."[8]

Tenner's analogy points to the limitations of using the body as a model for human-built systems. Although urban infrastructure can decay and perform suboptimally, it does not fall ill in the same ways organisms do. The disjunction between the built environment and human health is rich with historical examples; a theme of the history of technology is unintended consequences of technological applications. The widespread use of dichlorodiphenyltrichloroethane (DDT) addressed causes of malaria and other insect-borne diseases; it also produced severe reproductive health problems in several species of animals. Application of naturally occurring materials in industrial society has had unintended consequences. Using the mineral asbestos had benefits that eventually were superseded by risks. After numerous devastating fires, American cities in the mid-twentieth century sought to make buildings safer by recommending the use of asbestos as a flame retardant. By the mid-1970s, the federal government sought to contain a public health problem by banning the use of asbestos in buildings, as inhaled asbestos fibers were linked to mesothelioma and other incurable cancers in the residents of affected buildings, in the workers who maintained those buildings, and in the miners excavating the mineral for use. Whether synthesized by humans or extracted by nature, substances used to improve safety have often produced consequences visible on or in the human body.[9]

## The Body in Systems

If bodies are systems, bodies also exist within systems. As chapter 2 shows, developing systems was an important aspect of industrialization. Thus, scholars concerned with industrial systems are also concerned with the body's role within those systems. Thinking about how to achieve a more embodied historical practice, as Katherine Hayles urged, led to consideration of people's bodies as components of industrial systems. According to this way of thinking, industrial accidents are not just labor, class, and economic problems; they demand an "ecological approach to industrial health and safety."[10]

Embodied approaches to history extend beyond the factory to all landscapes of labor. Thinking about how human labor functions in nature involves more than simply thinking about human acts of ecological destruction. Instead, historians analyze the complex interactions between human labor, systems of industrial work, and nature.[11] If human labor operates within systems, it can function to provide ecosystem services; for example, fighting forest fires or otherwise maintaining forests. Thinking about how ecosystems and bodies are stressed by toxins such as DDT or radioactive waste situate the body within larger systems.[12]

Changes in dynamic human work environments are linked to observable changes in the bodies performing labor in them. Conevery Bolton Valenčius observed that the American frontier mentality that Thomas Jefferson and Alexis de Tocqueville articulated in the eighteenth and nineteenth century was reflected in the lived experiences of Americans on the Arkansas frontier. Working long hours in the sun, people who racially identified as white saw their skin darken, raising questions about the mutability of (constructed) racial categories and the role of the environment in shaping physical characteristics.[13]

The American experience also raises the issue of how technology and the environment interacted to shape forced labor systems. Forced labor in the form of slavery and indentured servitude subjected workers' bodies to violence and oppression in a range of subjugations that included torture, rape, and murder. Agricultural systems in the Americas relied on slave labor, with technological innovations such as the cotton gin actually helping to intensify slavery in the decades before mechanization and "free labor" became the dominant economic model in the United States. The historical experience of slaves belies technologically determinist arguments that the Industrial Revolution rendered slavery obsolete. Angela Lakwete has described the use of technologies to cultivate cotton—technologies that evolved over centuries in China, India, and the Ottoman Empire—and thus transform the work of slaves in the Americas during the nineteenth century.[14]

States have taken interest in developing ideal forms of the body. In one US example, the federal government established the Civilian Conservation Corps (CCC) during the Great Depression, in part to "build men out of boys." In this approach, human bodies worked on the wilderness as the wilderness worked on those bodies. The use of rural work camps was crucial in transforming urban boys in a more rugged, virtuous environment. In the structured work environments of the New Deal, young urban males were transported to rural areas, provided with tools to build structures and cut forests, and trained in regimented work that both strengthened their muscles and began to train them for the types of interactions with weapons, jeeps, and regimented time required in military operations.[15]

The CCC represents government policy that models a complex interaction between technology and the environment that benefits both. Wilderness was transformed into buildings, and sickly, corruptible youth were transformed into muscular, useful men fit for military service. This optimistic vision of nature and technology informed policies whereby hydroelectric dam were built on rivers. When Woody Guthrie celebrated these projects in songs like "Pastures of Plenty" and "Grand Coulee Dam" (the latter referring to one of the many dams the federal government built on Washington State's Columbia River during the 1930s), the projects represented not only technological progress but also the potential to employ construction workers and reap benefits for farmers, benefits that strengthened American society as well as a construction of masculinity that had been challenged by widespread unemployment. In this context, serving as cogs in a large technological system was crucial for the development of American men.

This emphasis on training is one aspect of the use of human bodies within larger technical systems. Although mechanization replaced human labor in some cases, the Industrial Revolution radically transformed human labor in many others. The rationalization of production sought by Frederick Winslow Taylor and Henry Ford transferred expertise from individual workers who developed skills over long periods of apprenticeship to engineers who designed systems in which new workers could be quickly integrated without extensive training. Which is not to say the new dimensions of work were easy; fatigue from repetitive tasks and the hazards involved in working with machinery posed new risks.

This transformation often involved new relations between people and the environment. Mining ores underground, for example, underwent dramatic change, as miners faced suffocation, damage to their lung tissue by particulate matter, and potential cave-ins. Similarly, meatpacking and mechanized food production (as discussed in the food systems chapter) exposed workers to threats of maiming in fields and slaughterhouses. As economies transitioned toward new and more-threatening work-

Laborers, mostly boys, at the Ewen Breaker of the Pennsylvania Coal Company in South Pittston, Pennsylvania, photographed by Lewis Hine in January 1911. Coal dust floats in the air and coats the faces of the boys and young men. Working environments like coal mines or processing facilities suggest the porous border between environments and human bodies. *Library of Congress, Prints and Photographs Division (LCCN 2018676220)*

place environments, literary representations of heightened workplace risk (such as that confronted by French miners in Émile Zola's 1885 novel, *Germinal*) captured readers' imaginations.[16]

Human bodies adjusted to industrial work in other ways. The advent of scientific management rationalized workplaces, necessitating a dependence on precise timekeeping. Individual occupations were transformed and often deskilled through division of labor. Production became less dependent on one worker with multiple skills than on several workers, each with particular tasks that required less training. New machinery, ranging from the assembly line in factories to vacuum cleaners and microwave ovens in households, transformed the ways humans performed work and, in turn, altered the bodies required for it.[17]

Work transforms workers. The repetitive tasks and risks of injury involved in waste trade work not only shaped workers but also ascribed physical identities to the people who performed it. In addition to risking injury and illness from the chaotic, unclassified discards they sorted, scrap metal dealers were, along with painters, plumbers, and other workers, vulnerable to poisoning from lead, a soft, pliable, widely useful, and very dangerous metal. Lead's use in paint, gasoline, and a variety of other industrial applications expanded its presence in the bones and neurological systems of people who came in contact with it.

In an example of historical scholarship as political advocacy, in the 1970s, Samuel P. Hays used the dental records of children to show how lead infiltrated their bodies, and he urged bans on its use in gasoline and paint. The illnesses and injuries produced by human labor in the presence of toxins shaped understandings of hazards that brought about new regimes for control and regulation of hazards, in the hope of protecting future bodies.[18]

Assembly lines required workers to perform repetitive tasks constantly over a defined workday. With repetitive work came risks of stress injuries, ranging from tendonitis to eye strain. Chronic injury and illness also manifested in respiratory illnesses, cancers, and toxic poisonings.

Acute injuries from workplace accidents have ranged from minor abrasions to traumatic death. As technologies have come to mediate our interactions with environments, a cultural shift has occurred in the ways in which people think about the relationship between the environment and bodies, emphasizing specific effects that environmental factors have on bodily health. Changes in industrial agriculture have affected agricultural workers' health; in this way, the industrialized landscape has reshaped the bodies of people who work in food systems.[19]

Linda Nash has used a similar model to explore how medical doctors, public health officials, engineers, and laypeople have understood the intertwined issues of environment, health, race, and disease since the late nineteenth century. The rise of industry produced a litany of new injuries and illnesses. From the 1910s through the 1930s, industrial hygiene investigators fashioned a professional culture that reshaped understandings of workplace health risks. In industrializing societies around the world, fields of professional knowledge developed lexicons and trained experts to define and treat illnesses borne of industrial production.[20]

### The Body as Sink

Analyses of the transformation of bodies by toxic environments also augment the literature on bodies injured in industrial work. The concept of pollution, discussed in chapter 3, is germane to human bodies transformed by natural and synthetic materials with toxic properties. A theme in this literature involves people in toxic environments adapting their physical behavior to survive dangerous landscapes.

Building on themes of resistance and struggle in feminist discourse, historians who focus on the burdens that polluted environments place on bodies depict affected people as active agents in navigating risk. Inhabitants understand the land as constitutive of who they are, and they use their embodied knowledge to resist bodily threats. This narrative approach breaks from a model of agentless victims subject to change imposed by more powerful forces within society and nature. It also challenges models that value institutionalized production of knowledge as dominant over vernacular understandings of nature, health, and risk.[21]

In some respects, embodied knowledge of local environments is a trait crucial to all of human history. Hunter-gatherers' knowledge of which habitats were safer was crucial to their survival, as was knowledge of where to find safe drinking water. Locating human waste away from drinking water was important to the success of societies centuries before the germ theory of disease transmission informed modern water sanitation.

In industrial society, the creation of new chemical compounds, production of new toxic wastes, and failure of aging infrastructures require new adaptations. The threats posed to air, land, and water in the cluster of petrochemical processing sites in Louisiana (USA) earned the area the moniker "Cancer Alley." Similarly toxic conditions characterize industrialized landscapes around the world, including aluminum smelters in Hungary and Norway, nuclear power facilities in Ukraine and Japan, and petrochemical plants in India and Saudi Arabia.

In such environments, residents learn the risks of the landscapes and adapt accordingly. The adaptations are themselves not without hazards, but the knowledge humans develop in responding to toxic environments resonates with the adaptations residents of Flint, Michigan, made in boiling water, using bottled water, and otherwise responding to the risks of bacterial contagion and lead poisoning following the 2014 decision by managers of the municipal government to alter the sourcing of residents' drinking water from the Detroit Water and Sewage Department's facilities to a poorly maintained system drawing water from the nearby Flint River. How and what vulnerable people do in the wake of changing environments have implications for how we interpret disasters, and also raise questions about how bodies mediate extreme environments and extreme disruptions in their environments.

The history of endocrine disruptors (chemicals that interfere with the body's endocrine, or hormone, systems, causing cancerous tumors, birth defects, and other developmental disorders) illuminates how industrial processes have affected biological systems. Nancy Langston's research on hormone disruption in the bodies of women who were prescribed the synthetic estrogen diethylstilbestrol (DES) in the mid-twentieth century mirrors Rachel Carson's account, in *Silent Spring*, of DDT disrupting animal reproduction in the wild. This ecological approach to human health, Langston argues, emphasizes that "the body is enmeshed in a web of relationships, not isolated within a castle whose threshold can only be breached by a sustained attack from the outside."[22]

The envirotechnical implications of hormone disruption destabilize our understanding of bodies as natural. If a "sustained attack" can transform reproductive health, and even "sex itself," stable biological categories come into question, and if the environment is vulnerable to natural as well as technological disruption, the concepts of nature and natural come into question as well.

Human interactions with radioactive material offer other approaches to this web of relationships. Uranium mining and the operation of facilities involved in producing nuclear energy for weapons and electricity for domestic consumption expose workers to radioactive material. And nearby residents cope with intentionally produced radioactive waste and unintentional fallout from disasters, such as the catastrophic events at the nuclear power plants in Chernobyl (Ukraine) and Fukushima (discussed in chapter 4). In radioactive environments, human laborers are transformed by exposure over time. In investigating the aftermath of plutonium production in Richland, Washington (USA), and Ozersk, Russia, Kate Brown discovered how residents interpreted the illnesses and pain they endured as results of radioactive poisoning. Interviews with residents who articulated their sensory ordeals allowed Brown to lift the veil of secrecy surrounding these sites of national security. In this way, the body becomes a text revealing the environmental damage wrought by nuclear policies.[23]

This approach to reading the body can be used to articulate experiences of environmental inequality and advocate for restorative justice. As is evident in chapter 4, on disasters, compelling envirotechnical histories analyze the systems that produce unequal burdens and risks, and challenge the assumption of technology's neutral role in society.

An important example is Love Canal, New York (USA), in the late 1970s. Despite observing a series of rare cancers, workers and residents of Love Canal struggled to show that their illnesses were the result of industrial poisoning in the wake of medical and regulatory assessments of their environment as safe. This history raises questions about the ability of expertise to grasp environmental illnesses, questions with relevance not only to the experience at Love Canal but for investigation of environmental inequalities around the world. Addressing the tensions over expertise in defining disease, pollution, and risk is one way in which scholars have produced context for the concerns of environmental justice activists.[24]

The example of Love Canal is instructive in approaching how illness is defined, who has the agency to define it, and whether environments have acceptably safe levels of pollution. The contested terrain between residents and trained professionals produces knowledge in the form of characterizations of safe versus unsafe environments. One of the most potent tools available to residents is the self-reporting of bodily ills; activists within the environmental justice movement have used it to advocate for remediation of polluted environments.

Self-reporting relies on interpretation of sensory information, reflecting the value of sensory history for contemporary environmental policy debates.[25] Richard Newman argues that students of environmental literature analyze the narratives of people in poisoned landscapes and understand how they articulate their experiences in "toxic autobiographies."[26] Such sources are used in analyses of pesticide exposure; ethnographic accounts of workers' experiences being poisoned are evident in the regulatory history of California agribusiness.[27]

Most of this chapter subsection discusses how industrial

substances used in nature may pollute the body. Ending the discussion here, however, runs the risk of confirming a stark divide between technology and nature that does not exist. Artificially mimicking or supplementing hormone production has an industrial component and is rooted in the work of the endocrine system. Organisms in nature also disrupt the human body, as any reader who has suffered from hay fever or pet allergies can attest. This disruption, too, has a history; human management of allergies, from classifying ragweed pollen as poison to developing chemical approaches to controlling allergic reactions (in the form of pills, sprays, and physician-applied shots), reveals how we have industrialized the often-adversarial relationships between humans and other organisms.

Finally, air, land, and water can produce dangerous, even fatal effects on human bodies. A sudden heat wave, for example, can kill people living in structures without adequate ventilation or cooling systems. Air-conditioning systems may actually worsen the heat island effect in cities, making the urban environment even worse for residents with more-limited resources. Climate changes may also affect disease vectors, further complicating the body's experiences with its surrounding environment.[28]

## The Body's Effects on the Environment

Much of the literature on how the human body relates to nature and technology focuses on effects on the body; yet the body itself has potential effects on the environment. Ellen Stroud's recent research into the environmental history of dead bodies notes that the industry centered on preparing and disposing corpses may poison the air, land, and water, as well as the bodies of the industry's workers. The manipulation of the human body using technology and chemistry—including implants of pacemakers, intrauterine devices (IUDs), dental prosthetics, antidepressants, chemotherapy, and other medicine—has consequences for the handling of human remains. In addition, cultural practices for laying dead bodies to rest may alter the environment.

Stroud identifies several potential poisons, including carcinogenic formaldehyde used in embalming, heavy metals found in cremation ashes, and the composition of human bodies made toxic through technological manipulation during our lives. Pacemakers, for example, can improve and extend life; in death, the lead in their batteries poses the risk of heavy-metal poisoning for other organisms in contact with the corpse. Stroud's analysis links the history of the body to discard studies, revealing that the methods we use to treat and inter dead humans are technological systems that have environmental effects.

Human bodies transform environments before those bodies die. In a broad sense, of course, everything human represents actions that bodies take that affect their environs. In a more specific sense, however, biological waste processes reveal that the ways we humans transform ourselves have direct bearing on our waterways and the organisms that rely on them. The converse of synthetic hormones affecting human health (Nancy Langston's focus) is the widespread use of hormones (including estrogen as birth control) in human bodies, which in turn leads to their emission through urine affecting watersheds proximate to major industrial metropolises.

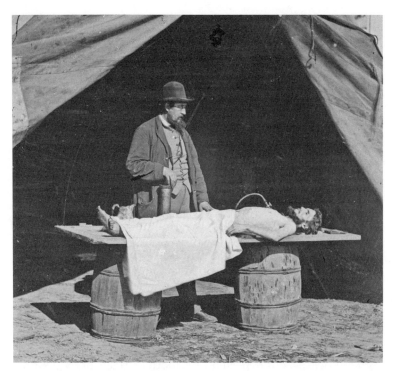

Dr. Richard Burr, US Army surgeon during the Civil War, demonstrates embalmment on a soldier. Extensive use of chemicals in the process shows how people can continue to have environmental effects even long after death. *Library of Congress, Prints and Photographs Division (LCCN 2018667145)*

Similarly, toxicologists have detected significant levels of pharmaceuticals ranging from antidepressants to opioids in waters surrounding large cities. Some of these compounds are caught in wastewater treatment facilities, but others pass through into the drinking water of area residents, representing a feedback loop of chemical alteration. Thus, bodies are both systems affected by outside phenomena and agents that impose change on ecosystems.[29]

———

The history of the body reveals the blurred boundaries between the environment and technology. Thinking about how human bodies mediate these blurred boundaries offers opportunities to think about how all living entities work within what we call the environment. Alert readers may notice repeated references to adaptation and evolution in this chapter; the wide variety of effects that bodies and artifacts in the human-built world have had on each other has inspired a model of evolutionary history.

Edmund Russell asked whether organisms themselves can be technology; his answer was "an emphatic yes."[30] Whereas Russell focused on nonhuman organisms such as dogs, the model of organisms as technology is also relevant to the study of the human body. Placing humans within the world of technology casts us as actors who transform and are transformed by technology. Through our food, our medicine, our work, and even our deaths, our experiences as both artifacts of nature and as subjects of technological manipulation reveal that the body is a fruitful subject for envirotechnical analysis.

Chapter Six

# Sensescapes

U niting insights from sensory studies and envirotech literature allows us to consider the sensory dimensions of both
the natural and the built environment, even as scholars have
questioned these categories and exposed the porous boundaries
between them. The term *sensescape*, which evokes the more
common *landscape*, suggests not just a physical site but also the
sensory dimensions of a space or a place. Anthropologist David
Howes defines sensescape as "the idea that the experience of the
environment, and of the other persons and things which inhabit
the environment, is produced by a particular mode of distinguishing, valuing and combining the senses in the culture under
study."[1] This chapter surveys some of the work, primarily by historians, that addresses emerging intersections between envirotech and the senses.[2]

Collectively, this work suggests the importance of industrialization to altering sensescapes, the way new materials alter the
temporality of sensescapes, the complex ways in which various
technologies both transformed sensory experiences and facilitated knowledge of the sensory environment, and the potential
to include nonhumans in sensory studies.

## Senses and Sensory Studies

To begin with, understanding sensescapes, especially in the past, poses real challenges, foremost for historical scholarship. One difficulty is the classic problem of the archive: not only whose sources historians can (or cannot) access, but also the extent to which the archive itself is a historical product inflected by power.[3] Comparatively few documents from eighteenth- and nineteenth-century industrial workers have survived, for instance, let alone ones detailing the acoustic or olfactory environment of the factory. Furthermore, even if historical sources do comment on sensescapes and how various groups perceived and experienced them, rarely do these materials fully capture how actors experienced them.

Some sources may provide, for example, visual representations of nonvisual phenomena. Sketches or photographs of urban rivers thick with stinky sewage help document pollution. Such images suggest what it may have been like to live or work there. Yet they do not preserve the actual "smellscape" of those places. Sensory studies scholars have discussed the methodological challenges of this research, as well as constraints presented by traditional academic publications, which prioritize textual and visual representations.[4] These problems are certainly shared by those working at the nexus of environmental history, the history of technology, and sensory history.

Despite these challenges, scholars to date have developed a number of substantial insights, starting with problematizing in several important ways any simplistic notion of the senses. Many of these insights extend to sensescapes. First, anthropologists

Tanneries along the river la Bièvre in Paris, photographed by Charles Marville (1813–1879) in 1862. Although the river appears fairly clean in this photo, the muck on tanners' aprons, boots, and equipment hints at the effluent dumped in waterways like la Bièvre. Still, photographs and other images fail to capture the full sensory effect of industries like tanneries—their smellscape. *Wikimedia Commons*

and historians have teased out the cultural specificity of the senses. They challenge the notion of a singular, self-evident, wholly biologically based understanding of the senses and sensory experiences. Hearing in West African Ewe communities, for example, does not map onto tidy Western divisions among the five senses. Instead, the experience of hearing is a more complex, embodied process that involves what Westerners believe are multiple, distinct senses. Indeed, the five senses may be remnants, albeit influential ones, of Aristotelian thought.[5]

Second, historical approaches foreground the specificity and contingency of sensescapes. For example, technical innovations such as aviation and new, synthetic materials transformed the soundscape and smellscape, respectively, in new ways. In some cases, established activities, when taken to an industrial scale, created new problems. Industrial waste disposal and processing of fish, for example, created smellscapes that exceeded historic norms, as well as the biological processes that had previously limited rank odors.[6]

Together, these cultural and historical contingencies demonstrate how both the senses and sensescapes are problematic, are situated, and are far from self-evident. People with synesthesia, for instance, experience the convergence of two or more senses simultaneously. Through empirical analyses, sensory studies scholars have thus questioned fundamental categories and assumptions, such as the number of and distinction among the senses—a third major contribution of sensory studies.

Fourth, environmental historians, environmental anthropologists, scholars in the interdisciplinary field of animal studies, and others working in what is called multispecies ethnography have the potential to unsettle sensescapes and sensory studies in valuable ways by questioning their typical anthropocentric orientation. After all, humans are not the only organisms or species who experience the environment through their senses. For one, the "same" senses—say, sight or smell—work differently in different species. Both humans and bees "see," yet they do so in distinct ways. In addition, the relative importance of each sense depends on the species. As dog owners know well, scent is meaningful to canines in ways it is not to their humans.[7] For bats, acoustical information and thus soundscapes are more important than sight. Furthermore, modifying the sensescapes around presumed human sensory subjects, often through technological interventions such as artificial lighting and freeways, may have unintended consequences for nonhumans.[8] Environmental humanists therefore offer valuable contributions to—but also critiques of—sensory studies by decentering humans. They invite us to consider how nonhumans experience diverse sensescapes in equally distinct ways.

Finally, although the history of sensescapes recalls established insights in environmental history and the history of technology regarding the transformative environmental effects of new technologies on nonhuman nature, analyzing sensescapes demonstrates how technology and the environment are not necessarily oppositional. Technologies can enable, mediate, transform, and constrain how we know and experience the natural world, including through our senses, in complex and sometimes-unexpected ways.

For those living in industrial cities, walking may be very different from Henry David Thoreau's experiences in the mid-nineteenth century. At the same time, for those with access to GPS (Global Positioning System), the technology can literally open up new vistas, as well as change where walkers go and what they may see.[9] Binoculars sharpen human vision over long distances, while hearing aids and parabolic microphones amplify sound. Transportation technologies—from the carriage to the railroad, car, and plane—mediate the geographical and temporal scales of the sensescapes through which we travel.[10] Technologies can thus become extensions of the human body, at times enabling, heightening, or augmenting, at other times altering, the senses. Envirotechnical approaches to sensescapes thus help complicate the boundaries between body, environment, and technology.

## Sound

With some of these broad considerations in mind, we now turn to several specific kinds of sensescapes, beginning with sound and hearing. We begin here in 1969, when composer and environmentalist R. Murray Schafer coined the term *soundscape*, which reflected both his disciplinary training and contemporary environmental politics.[11] Through the concept, Schafer reinforced but also challenged contemporary notions of the environment and environmentalism by extending these concerns to sound. *Soundscape* suggests the sensory, and specifically the acoustic, aspects of a physical space or place.

The field of sensory studies, and sound studies scholarship in particular, generally credits Schafer with helping draw attention to listening, acoustic environments, and their historical contingencies. The field of sound studies has problematized the Western tradition's visual orientation, a bias that permeates even mundane language and metaphor. In 2005, environmental historian Peter Coates sought to put this emergent research on the agenda of this field, arguing for "an environmental history of sound." Coates also suggested how sound studies research, despite Schafer, had not adequately addressed sound's environmental dimensions and implications. Environmental historians could thus bring new insights to the interdisciplinary field of sound studies.[12]

Integrating these literatures—sound studies, environmental history, and the history of technology—in a deeper way suggests an envirotechnical approach to sound. Indeed, Schafer's original concept *soundscape* is a generative idea for envirotech scholars. It calls attention to the sonic dimensions of both human and nonhuman nature. It also forges useful parallels between landscape and soundscape, including their shared natural-cultural dimensions. Lastly, it invites analysis of the technologies of hearing, knowing, and recording acoustic environments, and the aural experiences of both humans and nonhumans.

Industrialization helped remake diverse soundscapes, including those where people live, work, and play. As the chapter on industrialization suggests, traditional themes in this history include managerial control, labor conditions, shifting social hierarchies, and environmental effects. Yet attending to the acoustic context of industrial production illuminates how some workers also endured new kinds and levels of noise beginning in the eighteenth century. Loud, mechanized sounds became not just a daily working condition for employees in factories and mills but also a potential hazard. Progressive reformers in the late nineteenth and early twentieth centuries, and those who later studied them, focused on "improving" many aspects of the home and workplace—from child labor laws to technological systems such as water and sewer networks. Others were concerned about the environmental conditions of poor housing, dense neighborhoods, and poverty.

Examining the soundscapes of industrial modernity at work and home not only illustrates such historical patterns but also calls attention to less appreciated aspects of industrialization. Factories, highways, garbage and recycling facilities, and other industrial sites produced various forms of pollution. The contamination of air, soil, water, and human bodies is well known. Bringing an acoustic sensibility to these issues shows how they could also be sites of sound pollution.[13] Studying people's sonic environments suggests how sound affected health and well-being.[14] Yet race, class, and ethnicity shaped the aural experiences of different groups.

One key feature of industrialization was the growing reliance on fossil fuels, the internal combustion engine, and industrial transportation technologies. The dramatic consequences these factors have had for both built and natural environments over the past century have long been a mainstay in both environmental history and the history of technology. Less studied, however, are the acoustic dimensions of these interrelated technologies.

Considering sound opens up new perspectives on the history and politics of the built environment. The built environment of cities, suburbs, and homes entailed new aural experiences, particularly after 1945. The purported safety, comfort, and tranquility of suburbia in fact depended on automobiles and extensive road infrastructure—hardly quiet technologies. The post–World War II ideal of a single-family home, coiffed lawn, and yard suitable for playing children entailed considerable (male) labor for their upkeep. The seasonal routine of lawn mowers and leaf blowers regularly disturbed the supposed peace. The sounds of these suburban technologies also carried far beyond where they were used.[15] The actual soundscape of suburbia therefore conflicted with its ideals.[16] Moreover, in some contexts, the race, ethnicity, and class of landscape workers differentially exposed certain groups to harsh soundscapes that threatened their hearing.

Photographer Erik Calonius captured one aspect of subway construction in Washington, DC (USA), in 1972 as part of the series "DOCUMERICA: The Environmental Protection Agency's Program to Photographically Document Subjects of Environmental Concern, 1972–1977." The photo's original caption states that "noise from air compressors and jackhammers can cause irreversible hearing loss." Note that the woman walking by is covering her left ear. However, it's not clear from the photo whether the construction worker is wearing ear protection. Although a visual representation, this photo hints at the acoustic environment and soundscape of modern construction. *US National Archives and Records Administration (544968)*

The internal combustion engine and related technologies also had significant aural implications for the natural environment. On one hand, the automobile and roads facilitated access to national parks, wilderness, and other protected areas.[17] On the other hand, noise pollution extended far beyond the snaking paths of road networks in remote locales. Similarly, the geographic scale of aviation technologies and their acoustic reach created new challenges for formerly remote territories, now more readily accessible by air.

Growing tourism via helicopter and private plane in the Grand Canyon in the 1990s spurred the US National Park Service to begin investigating and addressing sound pollution in parks as part of its mission.[18] In 2005, scientist-activist Gordon Hempton established the "One Square Inch of Silence" sanctuary in Washington State, successfully lobbying for modifications to flight paths in and out of Sea-Tac Airport to protect a tiny piece of land on the Olympic Peninsula from sound pollution; but by doing so, these reforms shielded much of Olympic National Park from noise pollution. This project suggested how extensive air travel by the late twentieth century was affecting seemingly distant, isolated lands from the skies overhead.[19]

Evolving technologies, driven by political imperatives and global capital, have also had consequences for the ocean, an environment so vast it might seem beyond the possibility of anthropogenic effect.[20] Bioacoustic specialists studying the ocean have investigated how military submarines, sonar testing, and global shipping have impacted its whales and other charismatic megafauna.

As these examples suggest, the movement of sound waves

across various borders—from political to property boundaries—echoes a classic refrain in environmental history: the tensions between environmental and political borders. One of the field's founding insights is that environmental processes and non-human species do not necessarily respect political boundaries.[21] Indeed, transportation technologies that emerged over the twentieth century, including cars, trucks, and planes, amplified the geographic reach of industrial sound to unprecedented scales.

These technologies and their attendant sounds can traverse vast distances over land, in the sky, and through the ocean. They can also cross perceived borders, such as the sound of airplanes thousands of feet above shaping the soundscape of a terrestrial environment. New kinds of sounds and soundscapes therefore offer additional perspectives on large-scale anthropogenic environmental effects, or the Anthropocene. These cases begin to suggest how technologies can rework environmental boundaries, including soundscapes.

At the same time, technologies and techniques of listening are crucial to knowing soundscapes and materializing sound pollution, and thereby hint at relationships between technology and nonhuman nature that are more complex than technology serving as a powerful source of ecological change and degradation.[22] Scientific fields such as bioacoustics use various instruments to amplify human hearing, identify sound, record soundscapes, and visualize sound waves.[23] Placing recording devices on animals has brought into relief their acoustic environment and extended wildlife tracking beyond a focus on geographic location and mobility. Such devices thereby offer new understandings of the environment that encompass not only nonhumans' geography and range but also their soundscape.[24] These technologies of environmental knowledge making that record and visualize soundscapes have been critical to mobilizing concern about changing acoustic environments and, in turn, pushing for regulation and reform. In this sense, materializing soundscapes through various technologies was a crucial step toward materializing and legitimizing sound pollution as a problem in need of redress.

## Sight

As scholars in sound studies have long pointed out, sight and the visual have been privileged in the West since at least the Renaissance (the fifteenth and sixteenth centuries). Despite this bias toward the visual, we may examine sight more thoughtfully and reflectively by drawing on the analytical tools of sensory studies. Such research underscores the fact that even sight is not a transparent, universal way of seeing or visualizing. There is no view from "nowhere," what is sometimes called the God's-eye view or even "the god trick."[25] Rather, it too is culturally and historically situated. For instance, satellites orbiting Earth that image and visualize a global planet are inextricably connected to the Cold War and the military-industrial complex. Moreover, as glasses, binoculars, microscopes, telescopes, and satellites demonstrate, sight is often technologically mediated.[26]

Valuable work in the history of science, environmental history, and interdisciplinary studies of visualization has begun to explore how "the environment" is visualized and therefore conceptualized.[27] Among the most famous examples that both popular and scholarly histories have analyzed are various photographs of Earth from space, including *Earthrise* in 1968 and *Blue Marble* in 1972. Yet photographs are not unmediated representations of the "real world." In the case of *Earthrise*, astronauts saw the planet along one vertical edge of the moon; later, though, most reproductions of the photograph rotated Earth 90 degrees so that it looked as if the planet was rising over the moon's horizon. Rotating the photograph made it appear more like a traditional landscape painting, with the moon's gray, craggy surface in the foreground and the Earth appearing to rise vertically in the distance against the black cosmos. At times, such visual understandings were then materialized in the landscape. In envisioning the ideal environment for the German Autobahn (the national freeway known for having few speed limits), landscape architects aimed to create what they believed was a harmonious relationship between this large-scale technology system and the surrounding environment.[28]

Such critical studies of visualization, and of visualizing the environment in particular, show how technologies and technical practices shape how we define, think about, and know nonhuman nature.[29] For instance, satellite imaging technologies facilitated the visualization of the global or planetary environment.[30] These images illustrate Murphy's concept of regimes of (im)perceptibility.[31] Building on and extending Murphy's approach, we may also consider regimes of visual (im)perceptibility: that is, what is (not) made visible and thus literally (im)perceptible.[32]

Two landscapes of sight—lightscapes and nightscapes—illustrate how even visually oriented sensescapes, which are typically privileged in the West, can be taken for granted. Artificial light altered nighttime landscapes—or nightscapes—especially from the late nineteenth century with the advent of electric lighting. These trends intensified after World War II in much of the industrial urban world with the booms in suburbanization and consumption.

Artificial light at night extended day into night in homes, streets, neighborhoods, and public centers, offering practical comfort and ease—and, some believed, safety. In the late nineteenth and early twentieth centuries, engineers, marketers, writers, and others closely associated artificial lighting with powerful ideologies, including progress, modernity, and civilization.[33] The extension of light, and therefore day, into night illustrates how industrial capitalism has reshaped not only labor, production, and political economy, but also the nightscape as productivist work cultures have sought to stretch ever further the potential output of labor.[34]

Yet the ubiquity of artificial light in domestic, work, and public spaces has become so normalized and naturalized in most of the industrial, urban world, it garners little notice in the early twenty-first century. It has become infrastructure—until it fails.[35] Over time, gradual environmental changes, such as progressively brightening nighttime skies, have become so normalized that the current generation does not understand how it could be—and once was—otherwise.[36]

Indeed, because of changes since 1945, most people living in the global North in the early twenty-first century have never known truly dark nighttime skies.[37] Light pollution scientists have used various visualization techniques to dramatize artificial light at night in order to denaturalize the gradual brighten-

Extensive artificial light at night in Hong Kong in 2008. Light pollution is common in many urban industrial areas in North America, Europe, and Asia. Clouds at the top of the photograph reflect artificial light, thereby worsening light pollution. *"Hong Kong Night Skyline non-HDR."* Photo by Base64 via Wikimedia Commons. *Creative Commons Attribution–Share Alike 3.0 Unported license. No changes have been made.*

ing of nightscapes. By doing so, they seek to make this process literally and easily perceptible to the general public.

Just as the automobile, the plane, and other modern transportation technologies raise crucial questions about geographic scale, lightscapes and nightscapes centered on artificial light open up significant issues concerning temporal scale. Biologists and ecologists have highlighted the disjuncture between cultural change, such as the adoption of artificial lighting, and evolutionary time. They emphasize how human and nonhuman species are biologically adapted to a clear diurnal (day-night) cycle. However, the adoption of artificial light in urban industrial sites and the use of greater quantities of lighting have meant that some locales, such as Hong Kong, no longer have what is called true night. Just as scientists and scholars have stressed that fossil fuel use is in fundamental tension with geologic time scales because such use far outpaces replenishment (a key idea in chapter 2), light pollution scientists argue that expanding artificial light at night over the past century has ruptured biological processes resulting from generations of evolutionary change.[38] Put another way, rapid technological and cultural change has exceeded the typical pace of evolutionary change.

Although many scientists in this community use the term *light pollution* to describe artificial light at night, the phenomenon does not easily fit traditional understandings—whether lay, scientific, or activist—of pollution.[39] This mismatch is due in part to the sensory dimensions and sensescapes of light pollution. Artificial light at night is highly localized and temporary. It disappears by day and reappears by night. Yet the contingency of whether, when, and how it reappears depends on the schedule or whims of light users, individually and collectively. Artificial light at night is also a form of pollution that seems less tangible to the senses and less hazardous to the body. Unlike dark smoke rising from smokestacks or iridescent scum on the surface of water, it does not suggest overt pollution. It does not smell bad, unlike reeking sludge or noxious chemical fumes that scratch the throat and make eyes well up with tears.[40] Its seeming immateriality and apparent lack of direct, physical, immediately appreciable effects on the human or nonhuman body—in other words, the absence of sensory experience—likely help explain why light pollution is not yet on the agenda of mainstream environmentalism or environmental regulation.[41]

At the same time, nonhuman nature is not merely a passive

backdrop to technological change, which can bring about dramatic alterations to sensescapes such as lightscapes and nightscapes. Scientists studying artificial light at night, for instance, concluded that clouds increase light pollution by causing light waves to bounce back to Earth. The "same" amount of artificial lighting can therefore have greater (or less) effect, depending on whether nighttime skies happen to be clear or cloudy on a given night.[42] Particular nightscapes thus result from the precise conjuncture of technological and environmental phenomena, thereby challenging clear-cut boundaries between "natural" factors like weather and "human" factors like artificial lighting, and illustrating the fundamental premise of envirotech scholarship.

## Smell

Historian Alain Corbin was to the historical study of smell what R. Murray Schafer was to sound. Corbin's classic book, *The Foul and the Fragrant* (1986), helped lay the groundwork for what has become known as sensory history and sensory studies. Focusing on the olfactory dimensions of French history, Corbin aimed to trace how contemporaries perceived and understood smells, connecting these interpretations to larger debates over social order and political change in modern France.[43]

"Smelling the past" epitomizes some of the real challenges of doing sensory history. By their very nature, typical historical sources, dominated by texts and images, fail to preserve actual smells, whether fragrant or putrid. Instead, textual and visual representations mediate our knowledge of odors in the past.

Reading primary sources for long-ago scents, aromas, fragrances, and sometimes stenches can nonetheless provide some glimpses into history otherwise lost.

Photographs of Parisian streams that essentially became sewers and dumps suggest what it might have smelled like in the French capital.[44] Although writings by authors such as Charles Dickens are chiefly fictional, they can offer gritty, compelling accounts that dramatize banal, yet notable features of daily life. Nonfiction writers, such as Upton Sinclair, offer vivid descriptions of slaughterhouses that bring to life the smells and stenches, as well as sights, of the animal-industrial complex.[45]

Historians of technology, environmental historians, and those working at the intersection of these fields have developed rich analyses of technological systems such as water and sewer networks that transformed the natural and built environments. More attention could be paid to the olfactory aspects of these systems, what they attempted to replace, and how the resulting infrastructure altered smellscapes. These systems, and their precursors, were important to shaping the olfactory landscapes of premodern and modern eras.

Dolly Jørgensen has studied medieval sanitation techniques, usually seen as primitive and ineffective. Her research shows how such infrastructure was more effective than typically understood, indicating that such characterizations likely illustrate a postmedieval critique of the "dark ages" and attempt to differentiate "modern" approaches to urban infrastructural systems.[46]

Despite Jørgensen's caution, we should avoid idealizing earlier olfactory environments. Old industries such as tanneries required ample supplies of water and waste outlets. Neighbors and downstream users frequently criticized the industry for not just dumping animal carcasses and offal (guts of butchered animals) into streams and lakes and contaminating water supplies but also creating a stench surrounding the industry.[47] As urban populations grew and residents lived in former hinterlands, city residents complained about the foul smells (and sounds) of slaughterhouses now in their midst.[48] In response to these concerns, urban reformers around the turn of the twentieth century sought to modernize existing infrastructure in an attempt to improve the health and safety of city residents and to reduce some of the environmental consequences of dense human populations. Such environments included smellscapes.[49]

On one hand, then, industrialization, including expanded and modernized water, sewage, and other infrastructure, sought to clean up existing smellscapes. On the other hand, industrial processes also changed aromas in diverse settings—increasing odors, moving them to new areas, and sometimes bringing new smells and therefore smellscapes into being. Important innovations in chemistry and the chemical industry, and the development of synthetic goods, resulted in new materials, yielding new odors.

Over the twentieth century, synthetic smells such as chemical fertilizers replaced some biological odors such as animal manure and guano.[50] The material properties and qualities of these shifting forms of pollution mattered. Burning refuse in an era of plastics and other synthetic materials releases fumes that can be toxic and particulates that may not decompose for generations.[51]

Many of the odors and smellscapes discussed above are foul,

unclean, and potentially harmful. As we saw in chapter 3, on discards, anthropologist Mary Douglas has argued that "clean" and "dirty" odors are not neutral descriptions; she showed how cultural understandings of pollution are powerful forms of social (re)ordering. Carl Zimring's recent research has shown that elites perceived certain people and places in the United States as "clean" or "dirty." These associations were, however, mutually reinforcing, contributing to environmental racism. Exposure to and association with unpleasant odors thus not only reflects but also reinforces social inequalities.[52]

Contemporary garbage workers in the New York City Sanitation Division illustrate this point. These workers undertake the grimy, smelly labor of removing metropolitan trash. They are invisible to most city residents, and their work is imperceptible and devalued—until strikes or snowstorms stop garbage pickup.[53] In contrast, pleasant fragrances often imply purity, cleanliness, and health.[54]

In the late nineteenth and early twentieth centuries, such associations fostered back-to-nature movements as a way to escape the hazards of modern industrial life. According to this view, fresh air on farms and in camps and parks promised rejuvenation.[55] Douglas's theoretical framework and empirical studies like these suggest how scents and smellscapes are infused with cultural and political meanings. At the same time, they can obscure the conditions of their own making. The natural environment may seem pure and clean, yet such experiences rely on extensive transportation infrastructure and fossil fuels.[56]

Getting whiffs of fragrant aromas, whether sweetly scented

US Farm Security Administration photographer Russell Lee captured this "tester smelling cream to determine its freshness" at the Dairymen's Cooperative Creamery in Caldwell, Idaho, in 1941. It shows a reliance on smell and, specifically, olfactory expertise. *Library of Congress, Prints and Photographs Division (LCCN 2017789918)*

or malodorous, can certainly apply to the human sensory experience of smelling the world. Yet Chris Pearson has argued that we should also "sniff the past."[57] Pearson uses the verb *sniffing*— not *smelling*—to allude to canines. Dogs' enhanced sense of smell has made them valuable to humans for several reasons, including their ability to track animals, police suspects, and illicit substances.[58] Both hunters and police departments have used dogs as technologies for their olfactory expertise.[59] Non-humans can therefore help people navigate sensescapes in useful, valuable ways precisely because their sensory experience of smell is different from that of humans. Moreover, canines provide an important reminder that humans are not the only beings who smell or encounter olfactory environments.

## Taste

As with sound and smell, typical historical sources and archives make it difficult to understand the past experience of taste—a sensory experience and sensescape that has significant synergies with earlier chapters on food and the body. Nevertheless, analyzing specific foods reinforces the historically and culturally contingent nature of taste. Take sugar, for example. As Sidney Mintz showed in his classic study, *Sweetness and Power*, sugar was used more as a spice than as an ingredient before the large-scale cultivation of sugarcane in the plantation economies of the Atlantic world beginning in the seventeenth century.[60]

As the history of sugar suggests, tastes associated with certain culinary traditions and parts of the world have in fact been historically contingent. It may be hard to imagine Southeast Asian curries without hot peppers, Italian cooking without the tomato, or Irish meals without the potato, but none of these crops and ingredients were incorporated into Eurasian cuisines until at least the sixteenth century. "Classic" cuisines associated with certain cultures around the world were, in fact, not "native" or "natural."

As these examples demonstrate, prominent crops and culinary ingredients became possible thanks only to the intertwined processes of political, economic, technological, and ecological change, in these cases during European colonialization. Such cultural adaptation, creativity, and cosmopolitanism provide an important counternarrative to basic declensionist narratives. Nonetheless, such culinary and ecological shifts were firmly entangled with empire.

In addition, understanding of what constitutes taste is far from static. Four initial basic tastes—sweetness, sourness, saltiness, and bitterness—have been expanded to five with the addition of umami.[61] Yet both biology and cultural traditions also differ on these points. Sensitivity to bitterness is not uniform, while, according to Ayurvedic medicine, there are six foundational tastes. Furthermore, taste nicely illustrates the problematic boundaries among the senses that tidy categories simplify. Smell shapes taste, suggesting how multiple senses work in concert, even if they are understood as distinct.

The myriad ways in which foods represent interactions between nature and technology extend to their interactions as we eat and taste them. The act of eating, with its attendant cultural constructions, at once illuminates relations between humans and the environment, and social relations among people. Kyla Wazana Tompkins argues that the way people eat "produces political subjects by justifying the social discourses that create bodily meaning." Eating has long been a political act, linked to appetite, virtue, vice, race, and class inequality.[62] These patterns can be seen in recent trends toward "clean eating" and homemade, "locavore" cuisine, both of which reflect class privilege and usually create more work for women.[63]

Taste, then, is not timeless. The industrialization of taste resulted in part from wider transformations in agriculture, political economy, and technology.[64] As chapter 1, on food and food systems, demonstrates, the simplification, standardization, and commodification of agricultural crops such as corn have resulted in less diversity, growing reliance on a handful of intensely cultivated species, and ultimately standardized organisms.[65]

Industrialized food systems and the industrialization of taste have had complex effects. On one hand, they enhanced the sensory experiences of foodstuffs. Food producers used cultivation, chemistry, and genetic engineering to manipulate the appearance, taste, texture, and smell of foods. Processed food companies, ranging from Heinz to Dannon, developed research and development divisions to engineer "attractive" foods.

Multicolored candies and sodas in industrialized food systems are part of a carefully engineered practice to optimize taste and texture, visibly in the production of processed foods such as ketchup, pickles, and wine, and less conspicuously in meat, fruits, and vegetables. Industrialized kitchens test augmentations of food; successful ones lead to manipulations in fields and laboratories to produce the foods that elicit the most favorable sensory experiences.[66]

As part of the industrialization of taste, canned food produced by corporations (rather than by families) standardized recipes, adding set quantities of salt, preservatives, stabilizers, and other ingredients. The same brand of canned green beans or stewed tomatoes was supposed to taste the same. Industrially suited crops replaced local cultivars. Chemists, industrial researchers, and marketers—rather than personal taste or familial or cultural tradition—dictated what certain foods should taste like.

The movement of canned-food goods nationally and increasingly internationally enabled many families to eat according to mood, preference, and means, not according to the seasonal availability of crops or the labor of family members during previous seasons in canning, drying, fermenting, freezing, and otherwise processing foodstuffs for long-term consumption.[67] Overall, the development of the agro-industrial complex had important implications for food, cuisine, and taste.

Furthermore, industrial agriculture often gave less value to taste than to some other properties and qualities of food, such as the ability of crops to be harvested with mechanized equipment, transported long distances, or stored for long periods of time to facilitate the year-round availability of foodstuffs.[68] All of these goals became easier to attain over the twentieth century, as fossil fuels and long-distance transit facilitated and accelerated the circulation of foods.

At times, technological systems enabled certain people to transcend weather, seasonality, and other environmental constraints on food production and consumption. For example, in the late nineteenth century, elites in New Orleans began consuming icy cocktails in summer thanks to ice cut, stored, and transported from New England. Technologies of long-distance transit and climate control, including railroads, trucking, freeways, and refrigeration, enabled some historic boundaries of climate, geography, and seasonality to be ignored. However, such carbon-intensive systems brought fresh food from distant ecologies to certain markets with significant environmental and sometimes social consequences.[69]

Such technological changes and processes simultaneously entailed (re)shaping what, say, a tomato could or should taste like. At times, aesthetics supplanted actual taste; thus, the cultural sense of "taste" as an ability to discern and value—and its strong association with status—could become more important than literal taste. Pierre Bourdieu suggested how the two meanings of taste could become blurred.[70]

Early efforts at genetic engineering involved manipulating the freshness, firmness, and taste of certain foods, such as tomatoes, to extend their lives on supermarket shelves. Yet the convenience of longevity and year-round availability did not always prove commercially successful. The taste of the "Flavr Savr™" tomato—the first commercially grown, genetically engineered food, which was approved for human consumption in 1994—was so unpopular that the brand actually failed only three years later.[71] Industrializing taste has not always proved successful.

Examining these sensescapes opens up themes that resonate with and contribute to envirotech scholarship. Sensescapes illustrate complex relationships not only between nature and culture but also between technology and the environment. Industrialization played a crucial role in transforming diverse sensescapes—creating new geographical scales for sound, challenging biological processes that accommodated rank production and waste, and rupturing longer evolutionary processes through rapid technological and cultural change in artificial lighting. New materials generated new smellscapes at sites of production and disposal. Yet various technologies not only transformed sensory experiences but also facilitated knowledge of the sensory environment in new ways.

At the same time, examining sensescapes centered on and defined by Western definitions of the senses, as we have admittedly done here, risks reifying them. Combining these traditional categories and definitions of senses with "landscape" helps reveal the embodied experience of both humans and nonhumans in particular environments. Yet such an approach does miss the opportunity to interrogate those very categories. Furthermore, although humans have been the presumed subject of many senses and sensescapes, they are not the only ones.

# An Envirotechnical World

T he boundaries of social, technological, and environmental things and processes are often unclear and porous. Keeping this complexity in mind is useful as we live in and navigate a period of great upheaval. The term *Anthropocene*, applied to our current epoch, reflects the way human activities alter what we understand as the "natural" world. The planet's atmosphere surpassed 400 parts per million of carbon dioxide in 2016 for the first time in millions of years, a result of the centrality of carbon in the energy systems powering modern economies. Human-engineered compounds such as polychlorinated biphenyls (PCBs) have been found in the body tissue of polar bears in the Arctic, even though those animals live thousands of miles from where the chemicals were originally applied. In ways both intended and unintended, humans have reshaped our environs across the planet, and even beyond as "space junk" now clutters the lower atmosphere. Yet "humans," both past and present, are not equally responsible for these changes.[1]

The present effects of human interventions in the air, land, water, and life forms on Earth are vast and conspicuous. The themes we focus on in this book reveal interactions between technology and the environment in the past. In examining the histories of food, industrialization, discards, disasters, the body, and sensescapes, we often note the permeable boundaries between technology, the environment, and society. These histories also trace the creation and maintenance of envirotechnical systems, the prevalence of hybridity, the significance of biopolitics, and the continuing problem of environmental (in)justice. These themes are, in fact, interrelated. The development and maintenance of systems constitute an enduring theme in the history of technology, and in many respects, the themes of this book reflect the attention to the systems of production, transportation, and infrastructure that we associate with industrial society. However, our approach to systems assumes the permeable boundaries between what Thomas Parke Hughes called the "human-built world" and the land, air, water, and organisms of the Earth.[2]

Food systems have an envirotechnical history that dates back thousands of years before the advent of heavy industry and modernity. The agricultural revolution represented intricate technological means to extract nutrition from plants and animals, by diverting water, tilling soil, and cultivating particular organisms for human consumption. Industrialization then accelerated human manipulations of organisms and ecosystems. If the packaged cookies and sodas on supermarket shelves represented the most obviously artificial engineering of raw materials into edibles, they sit on a continuum with many of the meats, fruits, vegetables, grains, and beverages that have structured industrial society's diets. Historically, these foods and the systems that deliver them have allowed humans to multiply and thrive. But they have also brought about consequences not only for ecosystems and biodiversity but also for human health.

We urge readers not to see the Industrial Revolution as a complete break in the history of human interventions in our environments. Seeing the environment as both shaping and being shaped by industrialization challenges tidy categories and binaries such as nature and culture, as well as the environment and technology. Seeing the "technological" and "environmental" in industrialization as fundamentally entangled pushes us to see humanity as part of, and dependent upon, the so-called natural world—even as we simultaneously use, manage, and transform it, at times dramatically. These dependencies are not limited to more-obvious, immediate connections between people and the environment—growing food, logging in forests—but also encompass highly technologically mediated relationships, such as industrialization and computing.[3] The difference is that these technologies often mediate and obscure human reliance on the environment.[4]

This idea extends to the ways we pollute our environs. Many innovations have made the classification and management of the wastes that have accompanied all human societies more difficult and dangerous, especially as the scale and complexity of our disposal have grown. Some of this disposal is readily measurable; atmospheric scientists' detection that the atmosphere was taking on carbon led to social and political mobilization about this measurable threat.[5] The activist group 350.org (founded by author and environmentalist Bill McKibben and others in 2007) uses the measure of 350 parts per million of carbon dioxide in the atmosphere as the appropriate target level for reducing the effects of climate change. That global carbon emissions have continued to accelerate since 2007 indicates the extent to which the economic and political systems we have constructed are able to resist and discount compelling evidence that they produce catastrophic consequences.[6] Some have even argued that the term "climate *change*" whitewashes what is actually taking place. For instance, in May 2019, the prominent British newspaper the *Guardian* decided to alter the language it uses about the environment: "Instead of 'climate change' the preferred terms are 'climate emergency, crisis or breakdown,' and 'global heating' is favoured over 'global warming.'"[7]

While carbon can be easily quantified, other forms of disposal do not lend themselves to simple measurement. Ocean plastics are, as Max Liboiron argues, a "wicked problem" that endangers life both in the water and on the land. The scale and difficulty of this problem defy easy classification or solution. Pollution on this scale is neither out of sight nor out of mind, and presents a challenge to orderly societies' attempts to minimize the hazards of waste. Consideration of *what* and *who* is harmed by systems of disposal reveals issues of power and justice within nations, as well as across the planet. We encourage readers to consider how these processes relate to peoples and places in indigenous communities, the global South, and the post-Soviet world, and in landscapes of inequality in the United States and Western Europe.[8]

Issues of inequality and risk also inform the ways we define disasters. Scholarship on both historical and contemporary disasters highlights the politics and real-world implications of calling a disaster "natural," "technological," or "social." Designating a disaster as "natural" can remove or diminish whatever responsibility human actors and institutions may bear for the consequences of design, investment, and regulatory decisions that endanger people and nonhumans.

The power relations seen in discard studies and disaster studies demand interdisciplinary approaches. Readers of this volume's references may recognize scholars trained in anthropology, sociology, science studies, cultural studies, and environmental studies, as well as several subfields of history besides environmental history and the history of technology. While our approach draws upon multiple disciplines, it is rooted in the concerns and practices of history, enabling us to understand the historical contexts and contingencies of these relationships— the way they emerged in the past, how they help explain the present, and how they might be shaped for better possible futures.

The permeable boundaries we analyze in multiple cases in this book extend not only to the inputs of industrial production, and the externalities of discards, but also to ourselves. The historical experiences of people as workers within larger systems, as ecosystems for other organisms, and as sites of illness and sensory phenomena reveal not only that humans subject other organisms to envirotechnical manipulation, but that we subject ourselves as well. We authors wrote this book's manuscript by typing on computers using our hands. We reviewed our words on screens we viewed through prescription lenses that allowed us to see this manuscript with clarity (and, we hope, wisdom).

Thinking about how human bodies mediate these blurred boundaries offers opportunities to think about how all living entities work within what we call the environment. Placing humans within the world of technology casts us as actors who transform and are transformed by technology, often in ways that discipline us to function within larger systems of production and control. Envirotech history has important contributions to make to biopolitical studies of oppression, struggle, and resistance across time and geography. Through our food, our medicine, our work, and even our deaths, our experiences as both artifacts of nature and as subjects of technological manipulation reveal that the body is a fruitful subject for envirotechnical analysis.

Understanding the various ways our bodies interact with our environs requires analysis of the senses we use to understand the world. Sensescapes illustrate complex relationships not only between nature and culture but also between technology and the environment. Technologies transformed sensory experiences and facilitated knowledge of the sensory environment in new ways. While this book's words and images inevitably privilege the visual, we hope readers will not take for granted the ways in which we experience the world through other senses, or the ways these sensory experiences have changed over time, are culturally specific, and may differ by species.

Although this book's themes are expansive, the histories we have included here are far from comprehensive. We offer the foregoing chapters in the hope that readers will consider how the sensitivities and contingencies we discuss may inspire engaged historical work that tells the stories of peoples, places, and processes that are lightly covered or altogether absent from this volume. Within the global South and indigenous spaces across the planet, stories of how people navigate the porous boundaries of technology and environment—and how, in fact, diverse communities define and understand them—will give us better understandings of what we have built, what we have altered, how we have shaped the past, and how we might shape the future. We hope that by thinking envirotechnically, we can engage with the surrounding world—and with one another—in ways that are more sustainable and just.

# Teaching Resources

## Chapter 1. Food and Food Systems

### Recommended Readings

Cronon, William. *Changes in the Land: Indians, Colonists, and the Ecology of New England.* New York: Hill & Wang, 1983.

Crosby, Alfred W. *The Columbian Exchange: Biological and Cultural Consequences of 1492.* 30th anniversary edition. Westport, CT: Praeger, 2003.

Cushman, Gregory T. *Guano and the Opening of the Pacific World: A Global Ecological History.* New York: Cambridge University Press, 2013.

Gorman, Hugh S. *The Story of N: A Social History of the Nitrogen Cycle and the Challenge of Sustainability.* New Brunswick, NJ: Rutgers University Press, 2013.

Schrepfer, Susan, and Philip Scranton, eds. *Industrializing Organisms: Introducing Evolutionary History.* New York: Routledge, 2004.

### Websites

Anthropology of Food. http://aof.revues.org/.

The Food Museum. http://www.foodmuseum.com/.

History of Food & Agriculture. http://museum.agropolis.fr/english /pages/expos/fresque/la_fresque.htm.

Overview of the Global Food System (National Institutes of Health). https://www.ncbi.nlm.nih.gov/books/NBK114491/.

Films

*Food, Inc.* (2008)
*King Corn* (2007)
*The Natural History of the Chicken* (2000)
*Sustainable: A Documentary* (2016)

## Chapter 2. Industrialization

### Recommended Readings

Pritchard, Sara B., and Thomas Zeller. "The Nature of Industrialization." In *The Illusory Boundary: Environment and Technology in History*, edited by Stephen Cutcliffe and Martin Reuss, 69–100. Charlottesville: University of Virginia Press, 2010.

Schrepfer, Susan R., and Philip Scranton, eds. *Industrializing Organisms: Introducing Evolutionary History*. New York: Routledge, 2004.

Steinberg, Theodore L. "An Ecological Perspective on the Origins of Industrialization." *Environmental Review* 10, no. 4 (1986): 261–76.

Tarr, Joel A. *The Search for the Ultimate Sink: Urban Pollution in Historical Perspective*. Akron, OH: University of Akron Press, 1996.

### Websites

Industrial Revolution. https://sourcebooks.fordham.edu/mod/modsbook14.asp.

The 1911 Triangle Factory Fire. https://trianglefire.ilr.cornell.edu/.

Teaching with Primary Sources: "The Industrial Revolution in the United States." Library of Congress. http://www.loc.gov/teachers/classroommaterials/primarysourcesets/industrial-revolution/pdf/teacher_guide.pdf.

**Films**

*The Day the World Took Off* (2004)
*Germinal* (1993)
*Hard Times* (1994)
*The Luddites* (1988)

## Chapter 3. Discards
### Recommended Readings

MacBride, Samantha. *Recycling Reconsidered: The Present Failure and Future Promise of Environmental Action in the United States.* Cambridge, MA: MIT Press, 2011.

Melosi, Martin V. *The Sanitary City: Environmental Services in Urban America from Colonial Times to the Present.* Abridged Edition. Pittsburgh: University of Pittsburgh, 2008.

Royte, Elizabeth. *Garbage Land: On the Secret Trail of Trash.* New York: Little, Brown, 2005.

Strasser, Susan. *Waste and Want: A Social History of Trash.* New York: Metropolitan Books, 1999.

Tarr, Joel A. *The Search for the Ultimate Sink: Urban Pollution in Historical Perspective.* Akron, OH: University of Akron Press, 1996.

Zimring, Carl A., and William L. Rathje, eds. *Encyclopedia of Consumption and Waste: The Social Science of Garbage.* Thousand Oaks, CA: Sage, 2012.

### Websites

Basel Action Network. http://ban.org.
Center for Health, Environment, and Justice. http://chej.org.
Discard Studies. http://discardstudies.com/.
Zero Waste International Alliance. http://zwia.org.

### Films

*Scrappers* (2010)
*The Story of Stuff* (2007)
*Trashed* (2012)
*Waste Land* (2010)

## Chapter 4. Disasters
### Recommended Readings

Fortun, Kim, and Scott Frickel, "Making a Case for Disaster Science and Technology Studies." An STS Forum on the East Japan Disaster, accessed May 15, 2018. https://fukushimaforum.wordpress.com /online-forum-2/online-forum/making-a-case-for-disaster-science -and-technology-studies/.

Knowles, Scott Gabriel. "Learning from Disaster? The History of Technology and the Future of Disaster Research." *Technology and Culture* 55, no. 4 (2014): 773–84.

McPhee, John. *The Control of Nature.* New York: Farrar, Straus & Giroux, 1989.

Smith, Neil. "There's No Such Thing as a Natural Disaster." Understanding Katrina Essay Forum. New York: Social Science Research Council, 2006. Available at *Items: Insights from the Social Sciences,* https://items.ssrc.org/understanding-katrina/theres-no-such-thing -as-a-natural-disaster/.

Steinberg, Theodore. *Acts of God: The Unnatural History of Natural Disaster in America.* New York: Oxford University Press, 2006.

Special issue on Hurricane Katrina. *Social Studies of Science* 37, no. 1 (2007).

Special issue on the triple disaster at Fukushima Daiichi. *Environmental History* 17, no. 2 (2012).

## Websites

Disaster Research Center. http://drc.udel.edu/.

Disaster STS Network. http://disaster-sts-network.org.

STS Forum on the East Japan Disaster. https://fukushimaforum
.wordpress.com.

Superstorm Research Lab. http://superstormresearchlab.org/.

Teach 3.11. http://www.teach311.org/.

Understanding Katrina: Perspectives from the Social Sciences. http://
understandingkatrina.ssrc.org/.

## Films

*Blue Vinyl* (2002)

*Containment* (2015)

*One Night in Bhopal* (2004)

*Safe* (1995)

*The Spill* (2010)

*Trouble the Water* (2008)

*When the Levees Broke* (2006)

*Women of Fukushima* (2012)

## Chapter 5. Body
### Recommended Readings

Lacquer, Thomas Walter. *Making Sex: Body and Gender from the
Greeks to Freud.* Cambridge, MA: Harvard University Press,
1990.

Langston, Nancy. *Toxic Bodies: Hormone Disruptors and the Legacy
of DES.* New Haven, CT: Yale University Press, 2010.

Mitman, Gregg. *Breathing Space: How Allergies Shape Our Lives and
Landscapes.* New Haven, CT: Yale University Press, 2008.

Tenner, Edward. *Our Own Devices: The Past and Futures of Body
Technologies.* New York: Knopf, 2003.

Valenčius, Conevery Bolton. *The Health of the Country: How American*

*Settlers Understood Themselves and Their Land.* New York: Basic Books, 2002.

## Websites

Commonweal. http://www.commonweal.org/.

Consortium for History of Science, Technology, and Medicine. http://www.chstm.org/.

History of Science, Technology, and Medicine Database. https://hssonline.org/resources/hstm-database/.

Science and Environmental Health Network. http://www.sehn.org/.

## Films

*Blue Vinyl* (2002)

*Maquilapolis* (2006)

*Milk* (2015)

*Safe* (1995)

*Song of the Canary* (1979)

*What's Motivating Hayes* (2015)

## *Chapter 6. Sensescapes*
### Recommended Readings

Coates, Peter A. "The Strange Stillness of the Past: Toward an Environmental History of Sound and Noise." *Environmental History* 10, no. 4 (2005): 636–65.

Howes, David, ed. *Senses and Sensation: Critical and Primary Sources.* New York: Bloomsbury, 2018.

Howes, David. *Empire of the Senses: The Sensual Culture Reader.* New York: Berg, 2005.

Le Breton, David. *Sensing the World: An Anthropology of the Senses.* Translated by Carmen Ruschiensky. New York: Bloomsbury, 2017.

Smith, Mark M. *Sensing the Past: Seeing, Hearing, Smelling, Tasting, and Touching in History.* Berkeley: University of California Press, 2007.

**Websites**

Becoming Sensor in Sentient Worlds. https://becomingsensor.com/.
Centre for Sensory Studies. http://centreforsensorystudies.org/.
Megaprojects New Media. http://megaprojects.uwo.ca.
Sensory Studies. http://www.sensorystudies.org/.

**Films**

*Human Senses* (2003)
*Safe* (1995)

## Introduction. Technology and the Environment in History

1. An extensive literature has emerged in the environmental human-ities, cultural studies, and related fields centering on the concept of nature in the Western tradition. Less scholarship is dedicated to critical analysis of the concept of the environment. Many studies refer to the environment uncritically, as a self-evident scientific term. Work in the history of science and STS has begun to problematize this concept, show-ing how, for example, related terms such as ecosystem are embedded within and have resulted from the Cold War. Given the large literatures here, we are setting aside detailed debates over this terminology and its definitions, significance, and so forth. In this book, we refer to both the environment and nature, using the terms interchangeably so that we can maintain our primary focus on technology and the environment (or nature) in history. However, for a few key essays that help get into these issues, on "nature," see Williams, *Keywords*; Williams, *New Keywords*; Williams, "Ideas of Nature." On "ecosystem," see Martin, "Proving Grounds. On "environment," see Sutter, "The World with Us."

2. Much envirotech work to date has focused on the so-called West, as does our own research. The themes outlined here and in the greater envi-rotech literature will undoubtedly be complicated—even challenged—by studies of East and South Asia, Africa, Latin America, and the Middle East. Deeper engagement with sites in the socialist and communist worlds will also bring new insights. We hope that other scholars will take up these issues and write truly global envirotechnical histories. For example, the envirotech histories of East Asia group at EHESS (the

School for Advanced Studies in the Social Sciences, in Paris) constitute an important node in East Asian scholarship.

3. For instance, it is worth noting that "envirotech" presumes an initial separation between nature and technology. Not all cultural traditions assume nature and culture are so distinct. Thanks to Julia Adeney Thomas for pushing our thinking on this point. Her feedback originally responded to the following: Pritchard, "An Envirotechnical Disaster." Similarly, historians of technology have teased out the history of the term and idea. See Marx, "The Idea of 'Technology' and Postmodern Pessimism"; Kline, "Construing 'Technology' as 'Applied Science,'"; Schatzberg, "*Technik* Comes to America"; Marx, "Technology"; and Schatzberg, *Technology*. For a postcolonial critique of technology, see Mavhunga, *Transient Workspaces*. In addition, historical, cultural, and scholarly differences between *nature* and *the environment* merit more discussion. For one, *nature* and *the environment* can be both actors' and analysts' categories. There are also important historical and cultural questions here in terms of which concepts make sense within certain contexts. As mentioned, historians of technology have analyzed and historicized the term *technology*, thereby challenging the notion that it is an unproblematic category. Nonetheless, all scholars, including historians, need concepts and terms to organize and narrate their studies. And just as scholars of gender studying the medieval era, for instance, may study assumptions about and relationships between women and men even though contemporaries did not use the idea of gender, we can use nature and technology to interrogate the past. Yet, as these caveats suggest, we need to do so carefully and thoughtfully.

4. For a more inclusive definition of technology, see Nina E. Lerman, Ruth Oldenziel, and Arwen P. Mohun, "Introduction: Interrogating Boundaries," in Lerman, Oldenziel, and Mohun, *Gender and Technology*, 2. On "old" technologies, see Edgerton, *The Shock of the Old*.

5. Russell and Vinsel, "After Innovation, Turn to Maintenance"; Greene, *Horses at Work*; Edgerton, *The Shock of the Old*.

6. Hecht, *Being Nuclear*.

7. Worster, "Appendix: Doing Environmental History," in Worster, *The Ends of the Earth*, 289–307.

8. These review essays include Stine and Tarr, "At the Intersection of Histories"; Gorman and Mendelsohn, "Where Does Nature End and Culture Begin?"; Pritchard, "Joining Environmental History with Science and Technology Studies"; Pritchard, "Toward an Environmental History of Technology."

9. See, for example, the writings of John Stuart Mill.

10. On the Anthropocene, see Crutzen and Stoermer, "The 'Anthropocene'"; Waters et al., "The Anthropocene Is Functionally and Stratigraphically Distinct from the Holocene." On plastics, see Valentine, "Plastiglomerate, the Anthropocene's New Stone."

11. Weintraubjan, "More Female Sea Turtles Born as Temperatures Rise"; Richardson, *Sex Itself.*

12. The concepts of dialectical change and coproduction are useful here in terms of avoiding singular, linear causality.

13. See also Latour, *We Have Never Been Modern.*

14. Quoted in Langston, *Toxic Bodies*, 148.

15. On lead, see Hays, "Science and Values." On mercury, see Marran, "Contamination"; George, *Minamata*; George, "Fukushima in Light of Minamata"; and Walker, *Toxic Archipelago*. On radiation, see Petryna, *Life Exposed*; Hecht, "Nuclear Nomads"; Hecht, "The Work of Invisibility"; and Hecht, "Invisible Production and the Production of Invisibility." On endocrine disruptors, see Langston, *Toxic Bodies.*

16. For two accessible overviews, see Pollan, "Some of My Best Friends Are Germs"; Yong, *I Contain Multitudes.*

17. Hughes, *Networks of Power*. See also Hughes, *Human-Built World*, which introduces the concept "eco-technological environment."

18. On envirotechnical systems, see Pritchard, "Envirotech Methods"; LeCain, *Mass Destruction*; Pritchard, *Confluence*; and Pritchard, "An Envirotechnical Disaster."

19. Hybridity is also an important concept in postcolonial studies and theory. We are referencing instead ideas of hybridity developed largely

in science studies, the environmental humanities, and environmental history.

20. On hybridity, see White, *The Organic Machine*; Fiege, *Irrigated Eden*; Mitchell, *Rule of Experts*, 19–53; White, "From Wilderness to Hybrid Landscapes"; Reuss and Cutcliffe, *The Illusory Boundary*; Walker, *Toxic Archipelago*; and Sutter, "The World with Us."

21. Michel Foucault's publications are too numerous to list here, but a key starting point is his *Society Must Be Defended*.

22. Alatout, "Towards a Bio-territorial Conception of Power"; Agrawal, *Environmentality*.

23. The now-classic essay on the Anthropocene is Crutzen and Stoermer, "The 'Anthropocene.'" A more recent overview is Waters et al., "The Anthropocene Is Functionally and Stratigraphically Distinct from the Holocene." For just a few of the influential humanistic critiques of the Anthropocene and alternative framings, see Haraway, "Anthropocene, Capitalocene, Plantationocene, Chthulucene"; Haraway, *Staying with the Trouble*; Moore, *Anthropocene or Capitalocene?*; Moore, "The Capitalocene, Part I"; Moore, "The Capitalocene, Part II"; Hecht, "Interscalar Vehicles for an African Anthropocene"; Hecht, "The African Anthropocene"; and Yusoff, *A Billion Black Anthropocenes or None*.

## Chapter 1. Food and Food Systems

1. On extreme environments on Earth and elsewhere, see Olson, "Political Ecology in the Extreme"; and Pyne, *Between Two Fires*, and Pyne's many related books on the role of fire in American environmental history.

2. McNeill and McNeill, *The Human Web*.

3. McNeill and McNeill, *The Human Web*; Zohary and Hopf, *Domestication of Plants in the Old World*.

4. Diamond, *Guns, Germs, and Steel*.

5. Cronon, *Changes in the Land*; Krech, *The Ecological Indian*.

6. Schrepfer and Scranton, *Industrializing Organisms*; Reynolds,

*Stronger than a Hundred Men*; McShane and Tarr, *The Horse in the City*; Greene, *Horses at Work*.

7. Crosby, *The Columbian Exchange*; McCann, *Maize and Grace*; Ott, *Pumpkin*.

8. Cronon, *Changes in the Land*.

9. Mintz, *Sweetness and Power*.

10. Kurlansky, *Salt*; Gilbert and Reynolds, *Trading Tastes*; Abbot, *Sugar*; Finger, "Trading Spaces."

11. Cronon, *Nature's Metropolis*; Pacyga, *Slaughterhouse*.

12. Cronon, *Nature's Metropolis*.

13. Vileisis, "Are Tomatoes Natural?"

14. Bitting, *The Canning of Foods*; Freidberg, *Fresh*; Vileisis, "Are Tomatoes Natural?"

15. Cronon, *Nature's Metropolis*; Hamilton, *Trucking Country*.

16. Sawyer, *To Make a Spotless Orange*; Fitzgerald, *Every Farm a Factory*; Sackman, *Orange Empire*; Hutchings, "Consuming Nature."

17. Deutsch, *Building a Housewife's Paradise*.

18. Melosi, *Garbage in the Cities*; Elmore, *Citizen Coke*; Zimring, *Aluminum Upcycled*.

19. Pisani, *From the Family Farm to Agribusiness*; Worster, *Rivers of Empire*; Pfaffenberger, "The Harsh Facts of Hydraulics"; Shallat, *Structures in the Stream*; Fiege, *Irrigated Eden*; Josephson, *Industrialized Nature*.

20. Stoll, *The Fruits of Natural Advantage*; Wright, *A Fishery for Modern Times*; MacLachlan, *Kill and Chill*; Li, *Fighting Famine in North China*; Muscolino, *Fishing Wars and Environmental Change in Late Imperial and Modern China*.

21. Russell, "Introduction: The Garden in the Machine"; Harlan, *Crops and Man*; Kloppenburg, *First the Seed*; Fitzgerald, *The Business of Breeding*; Perkins, *Geopolitics and the Green Revolution*.

22. McCann, *Maize and Grace*; Gammage, *The Biggest Estate on Earth*; Moon, *The Plough That Broke the Steppes*.

23. Gorman, *The Story of N*; Gunderson and Hollings, *Panarchy*;

Werner and Newton, *Nitrogen Fixation in Agriculture, Forestry, Ecology, and the Environment*; Melillo, "The First Green Revolution"; Cushman, *Guano and the Opening of the Pacific World*.

24. Worster, *The Dust Bowl*; Worster, *Rivers of Empire*, 10 (quotation).

25. Sinclair, *The Jungle*; Young, *Pure Food*; Cronon, *Nature's Metropolis*; Washington, *Packing Them In*; Zimring and Bryson, "Infamous Past, Invisible Present."

26. Taylor, *Making Salmon*.

27. Russell, *War and Nature*; Anderson, "War on Weeds." Zachary J. S. Falck's discussion of urban and suburban efforts to classify and control weeds is also instructive of human attitudes toward undesired organisms. See Falck, *Weeds*.

28. Carson, *Silent Spring*; Russell, "The Strange Career of DDT."

29. Friedel, "American Bottles"; Smith-Howard, *Pure and Modern Milk*.

30. Cronon, *Nature's Metropolis*, 207.

31. Cohen, *A Consumers' Republic*; Cooper, "Microlessons."

32. Hogan, *Selling 'Em by the Sack*; Hamilton, *Trucking Country*; Josephson, "The Ocean's Hot Dog."

33. Sinclair, *The Jungle*; Pilcher, *The Sausage Rebellion*; Boyce, "'When Does It Stop Being Peanut Butter?'"

34. Bohme, *Toxic Injustice*.

35. Schrepfer and Scranton, *Industrializing Organisms*.

36. Russell, "Can Organisms Be Technology?" 257.

37. Russell, "Can Organisms Be Technology?" 257–58.

38. Martineau, *First Fruit*, 229, 233; Vileisis, "Are Tomatoes Natural?" 237–38.

39. Russell, "Are Organisms Technology?" 254; Paxson, *The Life of Cheese*.

40. Pollan, *The Omnivore's Dilemma*; Smil and Kobayashi, *Japan's Dietary Transition and Its Impacts*; Winson, *The Industrial Diet*.

41. Gorman, *The Story of N*, 162.

## Chapter 2. Industrialization

1. Pritchard and Zeller, "The Nature of Industrialization"; Pritchard, "Toward an Environmental History of Technology."

2. For a broad overview, see Pacey, *Technology in World Civilization*. On Britain, see Allen, *The British Industrial Revolution in Global Perspective*. On France, see Horn, *The Path Not Taken*. On the United States, see Licht, *Industrializing America*; Hindle and Lubar, *Engines of Change*. More recent work has emphasized the global dimensions of industrialization. See, for instance, Stearns, *The Industrial Revolution in World History*. Although more focused on the history of capitalism, Sven Beckert's *Empire of Cotton* can also be read through the lens of global industrialization. Similarly, Sidney Mintz's classic *Sweetness and Power* argues that plantation agriculture in the Caribbean offered a critical model for industrialization in Britain.

3. Karl Marx, *Capital*; Engels, *The Condition of the Working Class in England*. For overviews of US industrialization, see Licht, *Industrializing America*; and Hindle and Lubar, *Engines of Change*.

4. Berg, *The Age of Manufacturers*; Honeyman, *Women, Gender, and Industrialization in England*.

5. Toynbee, *Lectures on the Industrial Revolution in England*; Wilson, "Arnold Toynbee and the Industrial Revolution." For one example that complicates a radical rupture, see Shulman, *Coal and Empire*, which discusses the steamship and its continued reliance on sails and currents to reduce coal consumption.

6. Thorsheim, *Inventing Pollution*; Cavert, *The Smoke of London*.

7. Dickens, *Hard Times*; Gaskell, *North and South*; Sinclair, *The Jungle*.

8. Quoted in McNeill, *Something New under the Sun*, 59.

9. Mitchell, *Carbon Democracy*.

10. On the early use of coal, see Evelyn, *Fumifugium*. For influential early secondary literature, see Wrigley, *Continuity, Chance, and Change*.

11. Yan, *Reconstruction Designs of Lost Ancient Chinese Machinery*, 65–68; Oleson, *Oxford Handbook of Engineering and Technology in the Classical World*, 787–88.

12. Hindle and Lubar, *Engines of Change*, chapter 1.

13. McNeill, *Something New under the Sun*, 14.

14. McNeill, *Something New under the Sun*, 13–15.

15. Jones, *Routes of Power*, 34.

16. McNeill, *Something New under the Sun*, 13–15.

17. Howe, *Behind the Curve*.

18. The now-classic article is Crutzen and Stoermer, "The 'Anthropocene.'"

19. Smith, *Uneven Development*; Hecht, "The African Anthropocene."

20. Vaclav Smil has written extensively on the history and politics of energy. One useful starting point is his *Energy in World History*.

21. Nystrom, *Seeing Underground*.

22. Piper, "Subterranean Bodies"; Andrews, *Killing for Coal*.

23. Sandlos and Keeling, "Living with Zombie Mines"; Voyles, *Wastelanding*.

24. Andrews, *Killing for Coal*; LeCain, *Mass Destruction*; Curtis, *Gambling on Ore*; Leech, *The City That Ate Itself*.

25. Black, *Petrolia*; Frank, *Oil Empire*.

26. Valenčius, *The Lost History of the New Madrid Earthquakes*. Valenčius's current research project focuses on hydrofracking and seismology.

27. Josephson, *Industrialized Nature*.

28. Greene, *Horses at Work*; Shulman, *Coal and Empire*.

29. McGaw, *Most Wonderful Machine*; Sahlins, *Forest Rites*.

30. Lakwete, *Inventing the Cotton Gin*; Edelson, *Plantation Enterprise in Colonial South Carolina*.

31. Isenberg, *Destruction of the Bison*.

32. Cronon, *Nature's Metropolis*; Igler, *Industrial Cowboys*; Isenberg, *Destruction of the Bison*; Washington, *Packing Them In*, especially chapter 3.

33. Arch, "From Meat to Machine Oil."

34. McNeill, *Something New under the Sun*, 237–43; Dolin, *Leviathan*; Burnett, *The Sounding of the Whale*.

35. On the relationship between production and reproduction (and the centrality of the latter to the former), see Merchant, *Ecological Revolutions*.

36. Lowood, "The Calculating Forester"; Rajan, *Modernizing Nature*. On nature as technological, see Russell, "Introduction: The Garden in the Machine"; and Pritchard, *Confluence*, especially the introduction.

37. Worster, *Rivers of Empire*; Steinberg, *Nature Incorporated*; White, *The Organic Machine*; Cioc, *The Rhine*; Blackbourn, *The Conquest of Nature*; Pritchard, *Confluence*.

38. Fiege, *Irrigated Eden*.

39. Mitchell, "Can the Mosquito Speak?" in *Rule of Experts*, 19–53; Sutter, "Nature's Agents or Agents of Empire?"; Mukerji, *Impossible Engineering*; Carse, *Beyond the Big Ditch*.

40. See Martín, "Mechanization and 'Mexicanization.'" See also Schrepfer and Scranton, *Industrializing Organisms*; Boyd, "Making Meat"; Vileisis, "Are Tomatoes Natural?"; Horowitz, "Making the Chicken of Tomorrow."

41. Cronon, "The Trouble with Wilderness"; Spence, *Dispossessing Wilderness*; Neumann, *Imposing Wilderness*; Jacoby, *Crimes Against Nature*; Dowie, *Conservation Refugees*.

42. Purdue, "Is There a Chinese View of Technology and Nature?" 111.

43. See, for instance, Evelyn, *Fumifugium*.

44. Tarr, *The Search for the Ultimate Sink*; Melosi, *The Sanitary City*; Melosi, *Garbage in the Cities*; McShane and Tarr, *The Horse in the City*; Brantz, "On the Nature of Urban Growth"; Brantz, "Animal Bodies, Human Health, and the Reform of Slaughterhouses in Nineteenth-Century Berlin."

45. Brüggemeier, "A Nature Fit for Industry"; Uekötter, *The Age of Smoke*; Mauch and Zeller, *Rivers in History*.

46. Elkind, *Bay Cities and Water Politics*; Hamblin, *Poison in the Well*; Brown, *Plutopia*; Melosi, *The Sanitary City*.

47. Thorsheim, *Inventing Pollution*.

48. On the convenience of tall smelter towers, see LeCain, *Mass Destruction*.

49. Wirth, "The Trail Smelter Dispute"; Wirth, *Smelter Smoke in North America*; Disco and Kranakis, *Cosmopolitan Commons*; Frank, "The Air Cure Town."

50. Josephson, *Industrialized Nature*; Scott, *Seeing like a State*; Brown, "Gridded Lives"; Brown, *Plutopia*; Shapiro, *Mao's War against Nature*.

51. Tarr, *The Search for the Ultimate Sink*. For expert-led approaches, see Stradling, *Smokestacks and Progressives*. On bottom-up reformers, see Longhurst, *Citizen Environmentalists*. Michelle Murphy contrasts various approaches to reform within a single case in *Sick Building Syndrome and the Problem of Uncertainty*.

52. See the works cited above by Melosi, Stradling, and Tarr. See also Reid, *Paris Sewers and Sewermen*; Schneider, *Hybrid Nature*.

53. Steinberg, *Nature Incorporated*; Rosen, "The Role of Pollution Regulation and Litigation in the Development of the U.S. Meatpacking Industry"; Rosen, " 'Knowing' Industrial Pollution"; Rosen, "Businessmen against Pollution in Late Nineteenth Century Chicago."

54. This is a central premise of environmental justice; see chapter 5 in this book. See also McGurty, *Transforming Environmentalism*; Sze, *Noxious New York*; Pellow, *Resisting Global Toxics*; Pellow, *Garbage Wars*; and Pellow and Park, *The Silicon Valley of Dreams*.

55. McEvoy, "Working Environments."

56. Valenčius, *The Health of the Country*; Petryna, *Life Exposed*; Allen, *Uneasy Alchemy*; Mitman, Murphy, and Sellers, "Introduction: A Cloud over History," and the entire volume 19 of *Osiris*; Nash, *Inescapable Ecologies*; Mitman, *Breathing Space*; Roberts and Langston, "Toxic Bodies / Toxic Environments," and related articles in this issue of *Environmental History*; Langston, *Toxic Bodies*; Parr, *Sensing*

*Changes*; Parr, "Our Bodies and Our Histories of Technology and the Environment"; Maher, "Body Counts"; Walker, *Toxic Archipelago*; Hecht, "Nuclear Nomads"; Pritchard, "An Envirotechnical Disaster"; Brown, *Plutopia*; Keller, *Fatal Isolation*.

**57.** Such critiques parallel recent debates over the term *Anthropocene* and the ways in which it erases distinctions among human groups and their differential roles in climate change and anthropogenic environmental change. For arguments in favor of the term *Capitalocene*, see Moore, *Anthropocene or Capitalocene?*; Moore, "The Capitalocene, Part I"; Moore, "The Capitalocene, Part II." For one public, feminist critique, see Raworth, "Must the Anthropocene be a Manthropocene?"

**58.** On environmental justice, a key starting point is Bullard, *Dumping in Dixie*. For other important works in this area, see Sze, *Noxious New York*; Washington, *Packing Them In*; Pellow, *Resisting Global Toxics*; Pellow, *Garbage Wars*; Hurley, *Environmental Inequalities*. For an illustration of how gender matters, see Langston, *Toxic Bodies*; see also Vogel, *Is It Safe?* On lead specifically, see Hays, "Science and Values."

**59.** Cronon, "The Trouble with Wilderness"; Spence, *Dispossessing Wilderness*; Neumann, *Imposing Wilderness*; Jacoby, *Crimes Against Nature*; Dowie, *Conservation Refugees*; Powell, *Vanishing America*. On labor and the environment, see White, "'Are You an Environmentalist or Do You Work for a Living?'"; and Andrews, *Killing for Coal*.

**60.** Hecht, *Being Nuclear*, 204–5.

**61.** On whiteness as a form of racial and environmental privilege, see Murphy, *Sick Building Syndrome and the Problem of Uncertainty*, chapter 5; see also Zimring, *Clean and White*.

**62.** LeCain, *Mass Destruction*.

**63.** Wilkinson, "The English Industrial Revolution"; Wrigley, *Continuity, Chance, and Change*.

**64.** Pomeranz, *The Great Divergence*.

**65.** Peter C. Perdue argues that in Pomeranz's framework, "interaction, ecology, and contingency have replaced separation, civilizational dichotomies, and determinism." Perdue, "Perdue on Pomeranz."

66. Pritchard, *Confluence*.

67. Russell, *Evolutionary History*, especially chapter 9.

68. For an accessible introduction to gendered approaches to the history of technology, see Lerman, Oldenziel, and Mohun, *Gender and Technology*. On industrial use, see Kline and Pinch, "Users as Agents of Technological Change."

69. Latour, *Science in Action*.

70. Hughes, *Networks of Power*. For a summary and critique of Hughes, see Pritchard, "An Envirotechnical Disaster."

71. For several broad discussions, see Mitchell, *Rule of Experts*, especially chapter 1; Bennett, *Vibrant Matter*; Nash, "The Agency of Nature or the Nature of Agency?"; and Latour, *Reassembling the Social*.

72. Marx, *Capital*; Engels, *Condition of the Working Class in England*.

73. Marks, *The Origins of the Modern World*. Such tensions between (ecological) production and reproduction are central to Karl Marx's notion of the metabolic rift. See Schneider and McMichael, "Deepening, and Repairing, the Metabolic Rift." On the centrality of (biological) reproduction to (economic) production, see Merchant, *Ecological Revolutions*.

74. Gerber, *On the Home Front*; Brown, *Plutopia*; MacFarlane and Ewing, *Uncertainty Underground*; Hamblin, *Poison in the Well*; Galison and Moss, *Containment* (film).

75. A very early book in this vein is Cipolla, *Guns and Sails in the Early Phase of European Expansion*. Important later work includes Schivelbusch, *The Railway Journey*; and Shulman, *Coal and Empire*.

76. Mintz, *Sweetness and Power*; Russell et al., "The Nature of Power"; Finger, "Trading Spaces"; Cushman, *Guano and the Opening of the Pacific World*.

77. White, *The Organic Machine*; Fiege, *Irrigated Eden*; White, "From Wilderness to Hybrid Landscapes"; Reuss and Cutcliffe, *The Illusory Boundary*; Pritchard, *Confluence*; Schneider, *Hybrid Nature*; Sutter, "The World with Us."

## Chapter 3. Discards

1. Douglas, *Purity and Danger*.
2. Thompson, *Rubbish Theory*.
3. Lucsko, *Junkyards, Gearheads, and Rust*; Zimring, *Aluminum Upcycled*.
4. Tarr, *The Search for the Ultimate Sink*; Gorman, *Redefining Efficiency*; Sellers, *Hazards of the Job*; Stradling, *Smokestacks and Progressives*; Thorsheim, *Inventing Pollution*.
5. McNeill, *Something New under the Sun*; Hecht, *The Radiance of France*; Hecht, *Being Nuclear*.
6. Grossman, *High Tech Trash*; Gabrys, *Digital Rubbish*; Minter, *Junkyard Planet*.
7. Tarr, *The Search for the Ultimate Sink*.
8. Jørgensen, "Cooperative Sanitation"; Jørgensen, "Local Government Responses to Urban River Pollution in Late Medieval England"; Jørgensen, "What to Do with Waste?
9. Hamlin, "Edwin Chadwick and the Engineers."
10. Hamlin, "Edwin Chadwick and the Engineers"; Rosen, *A History of Public Health*.
11. Tarr, *The Search for the Ultimate Sink*; Melosi, *The Sanitary City*.
12. Tobey, *Technology as Freedom*; Ogle, *All the Modern Conveniences*; Rome, *The Bulldozer in the Countryside*; Everleigh, *Privies and Waterclosets*; Molotch and Norén, *Toilet*.
13. Cowan, *More Work for Mother*; Hoy, *Chasing Dirt*.
14. Cioc, *The Rhine*; Solzman, *The Chicago River*; Pritchard, *Confluence*.
15. Barber, *A House in the Sun*; Willis et al., *Energy Accounts*.
16. Schneider, *Hybrid Nature*; Langston, *Toxic Bodies*.
17. Tarr and Zimring, "The Struggle for Smoke Control in St. Louis."
18. Kaijser, "Under a Common Acid Sky."
19. Olson, "NEOSpace"; Rand, "Falling Cosmos."
20. McShane, *Down the Asphalt Path*; McShane and Tarr, *The Horse in the City*.

21. Melosi, *The Sanitary City*.

22. Gille, *From the Cult of Waste to the Trash Heap of History*; Stokes, Köster, and Sambrook, *The Business of Waste*.

23. Rajaram, Siddiqui, and Khan, *From Landfill Gas to Energy*.

24. Rathje and Murphy, *Rubbish!*

25. Reid, *Paris Sewers and Sewermen*.

26. Zimring, *Cash for Your Trash*; Perry, *Collecting Garbage*.

27. Zimring, "The Complex Environmental Legacy of the Automobile Shredder."

28. Corey, "King Garbage"; Melosi, *Garbage in the Cities*.

29. Nagle, *Picking Up*.

30. Medina, *The World's Scavengers*; Furniss, *Metaphors of Waste*.

31. Hurley, *Environmental Inequalities*; Washington, *Packing Them In*; McGurty, *Transforming Environmentalism*; Taylor, *Toxic Communities*; Zimring, *Clean and White*.

32. MacBride, *Recycling Reconsidered*.

33. Jørgensen, *Making a Green Machine*.

34. McDonough and Braungart, *Cradle to Cradle*; McDonough and Braungart, *The Upcycle*; Salehabadi, "The Scramble for Digital Waste in Berlin."

35. McGurty, *Transforming Environmentalism*; Voyles, *Wastelanding*.

36. Hill, "Garbage Here, Recycling There"; Pellow, *Resisting Global Toxics*; Stokes, Köster, and Sambrook, *The Business of Waste*; Gregson and Crang, "From Waste to Resource"; Reno, *Waste Away*.

37. MacBride, *Recycling Reconsidered*; Zimring, *Aluminum Upcycled*.

38. Liboiron, "Redefining Pollution."

## Chapter 4. Disasters

1. Useful review essays include Kelman, "Nature Bats Last"; Dyl, "Lessons from History"; and Knowles, "Learning from Disaster?" On the Lisbon earthquake, see Coen, *The Earthquake Observers*. On the Paris

flood, see Jackson, *Paris Under Water*. On Bhopal, see Fortun, *Advocacy after Bhopal*; Jasanoff, *Learning from Disaster*; Jasanoff, "Bhopal's Trials of Knowledge and Ignorance."

2. On 9/11, see Hecht, "Globalization meets Frankenstein?" On Hurricane Katrina, see Kelman, *A River and Its City*, as well as the 2006 reprint of Kelman's book for a new preface that addresses Hurricane Katrina; Colten, *An Unnatural Metropolis*; and Kelman, "Boundary Issues."

3. Research on Fukushima Daiichi has flourished since the triple disaster. See the appendix of this book for a number of invaluable teaching resources. Starting points for scholarly work include Pritchard, "An Envirotechnical Disaster"; and Hindmarsh, *Nuclear Disaster at Fukushima Daiichi*. For a historical perspective on earthquakes in Japan, see Clancey, *Earthquake Nation*.

4. McPhee, *The Control of Nature*; Davis, *Ecology of Fear*; Smith, "There's No Such Thing as a Natural Disaster."

5. Steinberg, *Acts of God*.

6. Kelman, *A River and Its City*; Colten, *Unnatural Metropolis*; Kelman, "Boundary Issues."

7. For perspectives on the evacuation experience, see Brigitte Steger, "'We were all in this together . . .'"

8. The categorization of disaster clearly matters for prevention, regulation, mitigation, and policy measures. Other scholars have pointed to the ways that *disaster* can be quite productive, even profitable; take, for instance, the insurance and reinsurance industries. See Jarzabkowski, Bednarek, and Spee, *Making a Market for Acts of God*.

9. On agnotology (or the production of ignorance), see Proctor and Schiebinger, *Agnotology*. On regimes of (im)perceptibility, see Murphy, *Sick Building Syndrome and the Problem of Uncertainty*. See also Hecht, "The Work of Invisibility"; and Hecht, "Invisible Production and the Production of Invisibility."

10. We might note here parallels with the great recession that began in 2008 and the recent boom in scholarship on the history of capitalism.

11. Fortun and Frickel, "Making a Case for Disaster Science and Technology Studies"; Knowles, "Learning from Disaster?"; Oreskes, "Why I Am a Presentist." See also Oreskes and Conway, *Merchants of Doubt*; and Oreskes and Conway, *The Collapse of Western Civilization*.

12. For compelling analyses of Chernobyl and its effects, see Petryna, *Life Exposed*; Brown, *Plutopia*; and Brown, *Manual for Survival*.

13. Nixon, *Slow Violence and the Environmentalism of the Poor*.

14. Knowles, "Learning from Disaster?" 779.

15. For a parallel argument about what *Fukushima* now symbolizes, see Pritchard, "An Envirotechnical Disaster."

16. In addition to Nixon, *Slow Violence and the Environmentalism of the Poor*, see also Berlant, "Slow Death (Sovereignty, Obesity, Lateral Agency)"; and Choi, "Anticipatory States."

17. Fortun, *Advocacy after Bhopal*; Jasanoff, *Learning from Disaster*; Carson, *Silent Spring*; Beamish, *Silent Spill*; Petryna, *Life Exposed*; Brown, *Plutopia*; MacFarlane and Ewing, *Uncertainty Underground*.

18. Fortun, cited in Knowles, "Learning from Disaster?" 778. Other works on asthma include Mitman, *Breathing Space*; Keirns, "Short of Breath"; Jackson, *Allergy*; and Jackson, *Asthma*. See also Aronowitz and Keirns, "Breath of Life."

19. For one example of such politics, see Redfield, *Life in Crisis*.

20. Huber, *Lifeblood*. STS literature on infrastructure is also useful here.

21. This critique is true for both historical actors and scholars who have not adequately questioned the definition of *disaster*.

22. Murphy, *Sick Building Syndrome and the Problem of Uncertainty*.

23. The sound metaphor of quiet or silence builds on Fortun (cited in Knowles, "Learning from Disaster?" 778) and Carson, *Silent Spring*. My term *quiet crises* is a play on *The Quiet Crisis*, the title of Stewart L. Udall's 1963 book. On (im)perceptibility, see Murphy, *Sick Building Syndrome and the Problem of Uncertainty*.

24. Langston, *Toxic Bodies*. See also the 2002 film directed by Daniel B. Gold and Judith Helfand, *Blue Vinyl*.

25. Pauly, "Anecdotes and the Shifting Baseline Syndrome of Fisheries"; Kahn, cited in Bogard, *The End of Night*, 310.

26. Bogard, *The End of Night*, 252. The internal quote is Kahn; the rest of the quote is Bogard.

27. Cinzano, Falchi, and Elvidge, "The First World Atlas of the Artificial Night Sky Brightness."

28. Valenčius, *The Lost History of the New Madrid Earthquakes*. See also Coen, "Witness to Disaster"; and Parrinello, *Fault Lines*.

29. For one illustration of competing narratives, see Keller, *Fatal Isolation*.

30. Marran, "Contamination." See also Hogan, *Hiroshima in History and Memory*; George, *Minamata*; George, "Fukushima in Light of Minamata"; and Walker, *Toxic Archipelago*.

31. Pritchard, "An Envirotechnical Disaster." On system, see Hughes, *Networks of Power*. On normal accidents, see Perrow, *Normal Accidents*. See also Perrow's more recent book, *The Next Catastrophe*.

32. On the idea of hybrid landscapes, and on the hybrid turn more broadly within environmental history, see Fiege, *Irrigated Eden*; White, "From Wilderness to Hybrid Landscapes."

33. Greg Bankoff, quoted in Knowles, "Learning from Disaster?" 776.

34. Penney, "Nuclear Nationalism and Fukushima."

35. Knowles, "Learning from Disaster?" 777. See also Pritchard, "An Envirotechnical Disaster."

36. Knowles, "Learning from Disaster?" 775–77.

37. Klein, *The Shock Doctrine*. See also Loewenstein, *Disaster Capitalism*.

38. Pritchard, *Confluence*, especially the conclusion.

39. Kahn, "Children's Affiliations with Nature."

40. Scott Frickel and M. Bess Vincent, "Hurricane Katrina, Contamination, and the Unintended Organization of Ignorance."

41. Beck, *World Risk Society*.

### Chapter 5. Body

1. Useful overviews of this subject include Parr, "Our Bodies and Our Histories of Technology and the Environment"; and Maher, "Body Counts."

2. Hughes, *Human-Built World*, 2.

3. Adelson, *Making Bodies, Making History*; Grosz, "Bodies and Knowledge," 31; Gowing, *Women, Touch, and Power in Seventeenth-Century England*; Canning, "The Body as Method?"

4. Lacquer, *Making Sex*; Butler, *Bodies That Matter*; Maines, *The Technology of Orgasm*.

5. Cavanagh, *Queering Bathrooms*; Davis, *Contesting Intersex*; Halberstam, *Trans\**.

6. Tenner, *Our Own Devices*; Laemmli, "A Case in Pointe."

7. Geddes, *Cities in Evolution*; Jacobs, *The Death and Life of Great American Cities*; Eisenman, "Frederick Law Olmsted, Green Infrastructure, and the Evolving City"; Wheeler and Beatley, *The Sustainable Urban Development Reader*.

8. Tenner, *Why Things Bite Back*, 8.

9. Maines, *Asbestos and Fire*; van Horssen, *A Town Called Asbestos*.

10. Hayles, *How We Became Posthuman*; McEvoy, "Working Environments."

11. White, "'Are You an Environmentalist or Do You Work for a Living?'"

12. Roberts and Langston, "Toxic Bodies / Toxic Environments."

13. Valenčius, *The Health of the Country*.

14. Lakwete, *Inventing the Cotton Gin*.

15. Suzik, "'Building Better Men'"; Maher, *Nature's New Deal*.

16. Zola, *Germinal*. See also Sinclair, *The Jungle*. For scholarly discussion of workplace risks, see Andrews, *Killing for Coal*; LeCain, *Mass Destruction*; Walker, *Toxic Archipelago*.

17. Cowan, *More Work for Mother*; Barca, "Laboring the Earth"; Otter et al., "Forum: Technology, Ecology, and Human Health since 1850."

18. Reid, *Paris Sewers and Sewermen*; Hays, "The Role of Values in Science and Policy"; Warren, *Brush With Death*; Zimring, "Dirty Work"; Jack and Steinhardt, "Atomic Anxiety and the Tooth Fairy."

19. Nash, "The Fruits of Ill-Health"; Sackman, "Nature's Workshop"; Mitman, Murphy, and Sellers, "Introduction: A Cloud over History," and the entire volume 19 of *Osiris*; Nash, *Inescapable Ecologies*.

20. Sellers, *Hazards of the Job*; Nash, *Inescapable Ecologies*; Walker, *Toxic Archipelago*; Sellers and Melling, *Dangerous Trade*.

21. Allen, "Narrating the Toxic Landscape in 'Cancer Alley,' Louisiana"; Allen, *Uneasy Alchemy*.

22. Langston, "The Retreat from Precaution"; Roberts and Langston, "Toxic Bodies / Toxic Environments"; Langston, *Toxic Bodies*.

23. Petryna, *Life Exposed*; Brown, *Manual for Survival*; Hecht, "Nuclear Nomads"; Pritchard, "An Envirotechnical Disaster"; Hecht, *Being Nuclear*; Brown, *Plutopia*.

24. Blum, *Love Canal Revisited*; Washington, Rosier, and Goodall, *Echoes from the Poisoned Well*. Washington also edits the journal *Environmental Justice*, which publishes case studies on the adverse and disparate health impact and environmental burdens that affect marginalized populations all over the world.

25. Parr, *Sensing Changes*.

26. Newman, "Darker Shades of Green."

27. Murphy, *Sick Building Syndrome and the Problem of Uncertainty*; Harrison, *Pesticide Drift and the Pursuit of Environmental Justice*.

28. Mitman, *Breathing Space*; Keller, *Fatal Isolation*.

29. Stroud, "From Six Feet Under the Field"; Stroud, "Dead Bodies in Harlem."

30. Russell, "Evolutionary History"; Russell, "Can Organisms be Technology?"

## Chapter 6. Sensescapes

1. Howes, *Empire of the Senses*, 143.

2. Anthropological literature is especially rich with respect to the senses. The list of pertinent works is too large to reproduce here.

3. Stoler, *Along the Archival Grain*; Blouin and Rosenberg, *Archives, Documentation, and Institutions of Social Memory*; Roque and Wagner, *Engaging Colonial Knowledge*.

4. For a useful overview, see Smith, *Sensing the Past*. See also Roeder, "Coming to Our Senses"; and Smith, "Still Coming to 'Our' Senses." For one illustration of the limits to traditional scholarly literary genres, see Joy Parr's multifaceted project, which includes her book *Sensing Changes* and her accompanying "Megaprojects" website, http:// megaprojects.uwo.ca/; and Jessica Van Horssen's multidimensional project, which includes her book *A Town Called Asbestos*, and the related graphic novel by Jessica van Horssen and Rhada-Prema McAllister, *Asbestos, PQ*.

5. Geurts, *Culture and the Senses*; Kuriyama, *The Expressiveness of the Body and the Divergence of Greek and Chinese Medicine*; Kuriyama, *The Imagination of the Body and the History of Bodily Experience*. Thank you to TJ Hinrichs and Rachel Prentice for sharing their insights here.

6. Chiang, *Shaping the Shoreline*; Chiang, "The Nose Knows."

7. Haraway, "Situated Knowledges"; Haraway, *The Companion Species Manifesto*; Haraway, *When Species Meet*; Haraway, *Staying with the Trouble*; Tsing, *The Mushroom at the End of the World*; Kirksey, *The Multispecies Salon*; Kirksey and Helmreich, "The Emergence of Multispecies Ethnography."

8. As Sara Pritchard's research on the science of light pollution reveals, scientists in this community note that they need to "see" like other species, not just humans. One ecologist memorably explained that she designs her studies so as to "see like a turtle."

9. Thoreau, "Walking" (1862); Jørgensen, "Walking with GPS."

10. Jørgensen, "The Armchair Traveler's Guide to Digital Environmental Humanities"; Schivelbusch, *The Railway Journey*.

11. Coates, "The Strange Stillness of the Past." For overviews of sound studies, see Sterne, *The Sound Studies Reader*; and Pinch and Bijsterveld, *The Oxford Handbook of Sound Studies*.

12. Coates, "The Strange Stillness of the Past." Overall, the history of science and technology and science studies have been more formative to early work in sound studies—and vice versa; environmental history has been less influential. Coates's essay offers many insightful analyses and suggestions as to how environmental historians might develop such an approach to sound. At the same time, he repeats many established arguments, such as unproblematically reproducing environmental knowledge-making techniques with respect to sound.

13. For two classic essays, see Bullard, "Overcoming Racism in Environmental Decision Making"; and Wenz, "Just Garbage."

14. Thompson, *The Soundscape of Modernity*; Parr, *Sensing Changes*.

15. Steinberg, *American Green*; Rome, *The Bulldozer in the Countryside*; Josephson, *Motorized Obsessions*; Yochim, *Yellowstone and the Snowmobile*.

16. Huber, *Lifeblood*.

17. Sutter, *Driven Wild*.

18. See some of the work undertaken by the Park Service's Division of Natural Sounds and Night Skies, at https://www.nps.gov/orgs/1050/index.htm.

19. On scientist-activists, see Frickel, *Chemical Consequences*. On this sanctuary, see Hempton and Grossman, *One Square Inch of Silence*.

20. Rozwadowski, *Fathoming the Ocean*.

21. Donald Worster, "Doing Environmental History," in Worster, *The Ends of the Earth*, 289–307. See also Fiege, "The Weedy West."

22. The term *materialization* is from Murphy, *Sick Building Syndrome and the Problem of Uncertainty*.

23. Bruyninckx, "Sound Sterile." See also Helmreich, "Underwater Music."

24. Benson, *Wired Wilderness.*

25. The classic essay here is Haraway, "Situated Knowledges," especially 581.

26. Otter, *The Victorian Eye.*

27. On problematizing the environment, see Sutter, "The World with Us." For works at the intersection of visual studies and environmental history, see Dunaway, *Natural Visions*; Dunaway, *Seeing Green*; Maher, "Shooting the Moon"; Pritchard, "The Trouble with Darkness."

28. Zeller, *Driving Germany.*

29. Cosgrove, "Contested Global Visions"; Cosgrove, *Apollo's Eye.*

30. Höhler, *Spaceship Earth in the Environmental Age*; Wormbs, "Eyes on the Ice."

31. Murphy, *Sick Building Syndrome and the Problem of Uncertainty.*

32. Pritchard, "The Trouble with Darkness."

33. Nye, *Electrifying America*; Nye, "The Transformation of American Urban Space." For other broad overviews, see Schivelbusch, *Disenchanted Night*; and Jakle, *City Lights*. For a popular history, see Brox, *Brilliant.*

34. Crary, *24/7*; Shaw, "Night as Fragmenting Frontier."

35. On failures of artificial-lighting infrastructure, see Nye, *When the Lights Went Out.*

36. Environmental psychologist Peter Kahn calls this phenomenon "environmental generational amnesia." Kahn, quoted in Bogard, *The End of Night.*

37. Cinzano, Falchi, and Elvidge, "The First World Atlas of the Artificial Night Sky Brightness."

38. Such cycles tend to be modeled on temperate zones, ignoring the fact that diurnal cycles in far northern and far southern latitudes are quite different. On the broader theme of evolutionary history, see Russell, "Evolutionary History"; and Russell, *Evolutionary History.*

39. On terminology and its implications, see Sara B. Pritchard, Erin McLaughlin, and Michelle Shin, "Describing Artificial Light at Night."

40. There are parallels here with, for instance, multiple chemical

sensitivity discussed by Murphy, in *Sick Building Syndrome and the Problem of Uncertainty*, chapter 6.

41. For an examination of the conceptualization and construction of light pollution as an environmental problem in France, see Challéat and Lapostolle, "(Ré)concilier éclairage urbain et environnement nocturne."

42. Kyba et al., "Red Is the New Black." Other factors such as snow, ice, particulates, aerosols, and ground cover also matter. For discussion of snow and artificial light at night in the polar North, see Pritchard, "Field Notes from the End of the World."

43. Corbin, *The Foul and the Fragrant*.

44. Reid, *Paris Sewers and Sewermen*.

45. On the animal-industrial complex, see Boyd, "Making Meat."

46. Jørgensen, "Cooperative Sanitation"; Jørgensen, "Modernity and Medieval Muck." See also Jørgensen, "The Medieval Sense of Smell, Stench, and Sanitation."

47. For discussion of the US urban context, see Melosi, *The Sanitary City*; and Melosi, *Garbage in the Cities*.

48. Brantz, "Animal Bodies, Human Health, and the Reform of Slaughterhouses in Nineteenth-Century Berlin"; Brantz, "On the Nature of Urban Growth."

49. Melosi, *The Sanitary City*; Melosi, *Garbage in the Cities*.

50. Gorman, *The Story of N*; Cushman, *Guano and the Opening of the Pacific World*.

51. Liboiron, "Redefining Pollution and Action"; Valentine, "Plasti-glomerate, the Anthropocene's New Stone."

52. Zimring, *Clean and White*.

53. Nagle, *Picking Up*.

54. Smith-Howard, *Pure and Modern Milk*. Smith-Howard's current work focuses on "cleanliness."

55. Schmitt, *Back to Nature*. Less focused on the environmental dimensions but relevant nonetheless is Downs, *Childhood in the Promised Land*.

56. Sutter, *Driven Wild*.

57. See Pearson, *Sniffing the Past—Dogs and History* (blog).

58. Pearson, "Between Instinct and Intelligence." Regarding companion species and multispecies ethnography, See the citations in note 7 of this chapter.

59. On nonhumans or nature as technology, see Schrepfer and Scranton, *Industrializing Organisms*; Russell, "Introduction: The Garden in the Machine"; Russell, "Can Organisms Be Technology?"

60. Mintz, *Sweetness and Power*.

61. Petrick, "The Asian Roots of Umami."

62. Tompkins, *Racial Indigestion*.

63. Matchar, "Is Michael Pollan a Sexist Pig?"

64. Petrick, "The Arbiters of Taste"; Petrick and Fitzgerald, "In Good Taste."

65. Fitzgerald, *The Business of Breeding*; Fitzgerald, *Every Farm a Factory*; Schrepfer and Scranton, *Industrializing Organisms*. Historians of science have amassed a rich literature on model organisms that could be used to help frame industrialized agriculture. See, for example, see Kohler, *Lords of the Fly*; and Rader, *Making Mice*; and Creager, *Life Atomic*.

66. Smith, *Pure Ketchup*; Boyd, "Making Meat"; Soluri, "Accounting for Taste"; Soluri, *Banana Cultures*; Petrick, "Arbiters of Taste"; Petrick, "'Like Ribbons of Green and Gold'"; Hannickel, *Empire of Vines*.

67. For an accessible overview of these issues, see Pollan, *The Omnivore's Dilemma*. Deborah Fitzgerald's current work focuses on the industrialization of food.

68. Martín, "Mechanization and 'Mexicanization'"; Hamilton, *Trucking Country*. For more-popular histories of these issues, see Pollan, *The Omnivore's Dilemma*; and Kingsolver, *Animal, Vegetable, Miracle*.

69. Freidberg, *Fresh*; Freidberg, *French Beans and Food Scares*.

70. Bourdieu, *Distinction*. Thank you to Marina Welker for this connection.

71. Martineau, *First Fruit*; Vileisis, "Are Tomatoes Natural?"

### Conclusion. An Envirotechnical World

1. Jones, "How the World Passed a Carbon Threshold and Why It Matters"; Haave et al., "Polychlorinated Biphenyls and Reproductive Hormones in Female Polar Bears at Svalbard"; Rand, "Falling Cosmos"; Hecht, "The African Anthropocene."

2. Hughes, *Human-Built World*.

3. Ensmenger, "The Environmental History of Computing."

4. White, "'Are You an Environmentalist or Do You Work for a Living?'"

5. Howe, *Behind the Curve*.

6. "About: History," *350.org*, accessed June 22, 2019, https://350.org/about/#history; Oreskes and Conway, *Merchants of Doubt*.

7. Carrington, "Why the Guardian Is Changing the Language It Uses about the Environment."

8. Liboiron, "Redefining Pollution."

# Bibliography

Abbot, Elizabeth. *Sugar: A Bittersweet History*. London: Duckworth Overlook, 2010.

Adelson, Leslie. *Making Bodies, Making History: Feminism and German Identity*. Lincoln: University of Nebraska Press, 1993.

Agrawal, Arun. *Environmentality: Technologies of Government and the Making of Subjects*. Durham, NC: Duke University Press, 2005.

Alatout, Samer. "Towards a Bio-territorial Conception of Power: Territory, Population, and Environmental Narratives in Palestine and Israel." *Political Geography* 25, no. 6 (2006): 601–21.

Allen, Barbara L. "Narrating the Toxic Landscape in 'Cancer Alley,' Louisiana." In *Technologies of Landscape: From Reaping to Recycling*, edited by David Nye, 187–203. Amherst: University of Massachusetts Press, 1999.

Allen, Barbara L. *Uneasy Alchemy: Citizens and Experts in Louisiana's Chemical Corridor Disputes*. Cambridge, MA: MIT Press, 2003.

Allen, Robert C. *The British Industrial Revolution in Global Perspective*. Cambridge: Cambridge University Press, 2009.

Anderson, J. L. "War on Weeds: Iowa Farmers and Growth Regulator Herbicides." *Technology and Culture* 46, no. 4 (2005): 719–44.

Andrews, Thomas G. *Killing for Coal: America's Deadliest Labor War*. Cambridge, MA: Harvard University Press, 2008.

Arch, Jakobina. "From Meat to Machine Oil: The Nineteenth-Century Development of Whaling in Wakayama." In *Japan at Nature's Edge: The Environmental Context of a Global Power*, edited by Ian Jared Miller, Julia Adeney Thomas, and Brett L. Walker, 39–55. Honolulu: University of Hawai'i Press, 2013.

Aronowitz, Robert, and Carla Keirns. "Breath of Life: Asthma in Historical Perspective." Online component of an exhibition at the National Library of Medicine, Bethesda, Maryland, March 1999–March 2001. https://www.nlm.nih.gov/archive/20120918/hmd/breath/breathhome.html.

Barber, Daniel A. *A House in the Sun: Modern Architecture and Energy in the Cold War.* New York: Oxford University Press, 2016.

Barca, Stefania. "Laboring the Earth: Transnational Reflections on the Environmental History of Work." *Environmental History* 19, no. 1 (2014): 3–27.

Beamish, Thomas D. *Silent Spill: The Organization of an Industrial Crisis.* Cambridge, MA: MIT Press, 2002.

Beck, Ulrich. *World Risk Society.* Cambridge, UK: Polity, 1999.

Beckert, Sven. *Empire of Cotton: A Global History.* New York: Knopf, 2014.

Bennett, Jane. *Vibrant Matter: A Political Ecology of Things.* Durham, NC: Duke University Press, 2010.

Benson, Etienne. *Wired Wilderness: Technologies of Tracking and the Making of Modern Wildlife.* Baltimore: Johns Hopkins University Press, 2010.

Berg, Maxine. *The Age of Manufacturers, 1700–1820: Industry, Innovation, and Work in Britain.* New York: Routledge, 1994.

Berlant, Laurent. "Slow Death (Sovereignty, Obesity, Lateral Agency)." *Critical Inquiry* 33, no. 4 (2007): 754–80.

Bitting, Avrill W. *The Canning of Foods: A Description of Methods Followed in Commercial Canning.* Washington, DC: Government Printing Office, 1912.

Black, Brian. *Petrolia: The Landscape of American's First Oil Boom.* Baltimore: Johns Hopkins University Press, 2000.

Blackbourn, David. *The Conquest of Nature: Water, Landscape, and the Making of Modern Germany.* New York: W. W. Norton, 2006.

Blouin, Francis X., Jr., and William G. Rosenberg, eds. *Archives, Documentation, and Institutions of Social Memory: Essays from the Sawyer Seminar.* Ann Arbor: University of Michigan Press, 2006.

Blum, Elizabeth D. *Love Canal Revisited: Race, Class, and Gender in Environmental Activism.* Lawrence: University Press of Kansas, 2008.

Bogard, Paul. *The End of Night: Searching for Natural Darkness in an Age of Artificial Light.* New York: Little, Brown, 2013.

Bohme, Susanna Rankin. *Toxic Injustice: A Transnational History of Exposure and Struggle.* Berkeley: University of California Press, 2015.

Bourdieu, Pierre. *Distinction: A Social Critique of the Judgment of Taste.* Translated by Richard Nice. Cambridge, MA: Harvard University Press, 1984.

Boyce, Angie M. "'When Does It Stop Being Peanut Butter?': FDA Food Standards of Identity, Ruth Desmond, and the Shifting Politics of Consumer Activism, 1960s–1970s." *Technology and Culture* 57, no. 1 (2016): 54–79.

Boyd, William. "Making Meat: Science, Technology, and American Poultry Production." *Technology and Culture* 42, no. 4 (2001): 631–64.

Brantz, Dorothee. "Animal Bodies, Human Health, and the Reform of Slaughterhouses in Nineteenth-Century Berlin." *Food and History* 3, no. 2 (2005): 193–215.

Brantz, Dorothee. "On the Nature of Urban Growth: Building Abattoirs in 19th-Century Paris and Chicago." In *Cahiers Parisiens*, no. 5, edited by Jan E. Goldstein, 17–30. Paris: University of Chicago Center, 2010.

Brown, Kate. "Gridded Lives: Why Kazakhstan and Montana Are Nearly the Same Place." *American Historical Review* 106, no. 1 (2001): 17–48.

Brown, Kate. *Manual for Survival: A Chernobyl Guide to the Future.* New York: W. W. Norton, 2019.

Brown, Kate. *Plutopia: Nuclear Families, Atomic Cities, and the Great*

*Soviet and American Plutonium Disasters.* Oxford: Oxford University Press, 2013.

Brox, Jane. *Brilliant: The Evolution of Artificial Light.* Boston: Houghton Mifflin Harcourt, 2010.

Brüggemeier, Franz-Josef. "A Nature Fit for Industry: The Environmental History of the Ruhr Basin, 1840–1990." *Environmental History Review* 18, no. 1 (1994): 35–54.

Bruyninckx, Joeri. "Sound Sterile: Making Scientific Field Recordings in Ornithology." In *The Oxford Handbook of Sound Studies,* edited by Trevor Pinch and Karin Bijsterveld, 127–50. New York: Oxford University Press, 2012.

Bullard, Robert D. *Dumping in Dixie: Race, Class, and Environmental Quality.* Boulder, CO: Westview Press, 2000.

Bullard, Robert D. "Overcoming Racism in Environmental Decision Making." *Environment: Science and Policy for Sustainable Development* 36, no. 4 (1994): 10–44.

Burnett, D. Graham. *The Sounding of the Whale: Science and Cetaceans in the Twentieth Century.* Chicago: University of Chicago Press, 2012.

Butler, Judith. *Bodies That Matter: Discursive Limits of Sex.* New York: Routledge, 1993.

Canning, Kathleen. "The Body as Method? Reflections on the Place of the Body in Gender History." In *Gender and History in Practice: Historical Perspectives on Bodies, Class, and Citizenship,* 499–513. Ithaca, NY: Cornell University Press, 2005.

Carrington, Damian. "Why the Guardian Is Changing the Language It Uses about the Environment." *Guardian,* May 17, 2019. https://www.theguardian.com/environment/2019/may/17/why-the-guardian-is-changing-the-language-it-uses-about-the-environment.

Carse, Ashley. *Beyond the Big Ditch: Politics, Ecology, and Infrastructure at the Panama Canal.* Cambridge, MA: MIT Press, 2014.

Carson, Rachel. *Silent Spring.* Boston: Houghton Mifflin, 1962.

Cavanagh, Sheila L. *Queering Bathrooms: Gender, Sexuality, and the Hygienic Imagination.* Toronto: University of Toronto Press, 2010.

Cavert, William M. *The Smoke of London: Energy and Environment in the Early Modern City*. Cambridge: Cambridge University Press, 2016.

Challéat, Samuel, and Danny Lapostolle. "(Ré)concilier éclairage urbain et environnement nocturne: les enjeux d'une controverse sociotechnique." *Natures Sciences Sociétés* 22, no. 4 (2014): 317–28.

Chiang, Connie Y. "The Nose Knows: The Sense of Smell in American History." *Journal of American History* 95, no. 2 (2008): 405–16.

Chiang, Connie Y. *Shaping the Shoreline: Fisheries and Tourism on the Monterey Coast*. Seattle: University of Washington Press, 2008.

Choi, Vivian Y. "Anticipatory States: Tsunami, War, and Insecurity in Sri Lanka." *Cultural Anthropology* 30, no. 2 (2015): 286–309.

Cinzano, Pierantonio, Fabio Falchi, and Christopher D. Elvidge. "The First World Atlas of the Artificial Night Sky Brightness." *Monthly Notices of the Royal Astronomical Society* 328 (2001): 689–707.

Cioc, Mark. *The Rhine: An Eco-biography, 1815–2000*. Seattle: University of Washington Press, 2002.

Cipolla, Carlo M. *Guns and Sails in the Early Phase of European Expansion, 1400–1700*. London: Collins, 1965.

Clancey, Gregory K. *Earthquake Nation: The Cultural Politics of Japanese Seismicity, 1868–1930*. Berkeley: University of California Press, 2006.

Coates, Peter A. "The Strange Stillness of the Past: Toward an Environmental History of Sound and Noise." *Environmental History* 10, no. 4 (2005): 636–65.

Coen, Deborah R. *The Earthquake Observers: Disaster Science from Lisbon to Richter*. Chicago: University of Chicago Press, 2013.

Coen, Deborah R. "Witness to Disaster: Earthquakes and Expertise in Comparative Perspective." *Science in Context* 25, no. 1 (2012): 1–15.

Cohen, Lizabeth. *A Consumers' Republic: The Politics of Mass Consumption in Postwar America*. New York: Knopf, 2003.

Colten, Craig E. *An Unnatural Metropolis: Wresting New Orleans from Nature*. Baton Rouge: Louisiana State University Press, 2005.

Cooper, Ken. "Microlessons: Toward a History of Information-Age Cuisine." *Technology and Culture* 56, no. 3 (2015): 579–609.

Corbin, Alain. *The Foul and the Fragrant: Odor and the French Social Imagination.* Cambridge, MA: Harvard University Press, 1986.

Corey, Steven Hunt. "King Garbage: A History of Solid Waste Management in New York City, 1881–1940." PhD dissertation, New York University, 1994.

Cosgrove, Denis. *Apollo's Eye: A Cartographic Genealogy of the Earth in the Western Imagination.* Baltimore: Johns Hopkins University Press, 2003.

Cosgrove, Denis. "Contested Global Visions: *One-World, Whole-Earth,* and the Apollo Space Photographs." *Annals of the Association of American Geographers* 84, no. 2 (1994): 270–94.

Cowan, Ruth Schwartz. *More Work for Mother: The Ironies of Household Technology from the Open Hearth to the Microwave.* New York: Basic Books, 1983.

Crary, Jonathan. *24/7: Late Capitalism and the Ends of Sleep.* New York: Verso, 2013.

Creager, Angela N. H. *Life Atomic: A History of Radioisotopes in Science and Medicine.* Chicago: University of Chicago Press, 2013.

Cronon, William. *Changes in the Land: Indians, Colonists, and the Ecology of New England.* New York: Hill & Wang, 1983.

Cronon, William. *Nature's Metropolis: Chicago and the Great West.* New York: W. W. Norton, 1991.

Cronon, William. "The Trouble with Wilderness; or, Getting Back to the Wrong Nature." In *Uncommon Ground: Rethinking the Human Place in Nature,* edited by William Cronon, 69–90. New York: W. W. Norton, 1996.

Crosby, Alfred W., Jr. *The Columbian Exchange: Biological and Cultural Consequences of 1492.* 30th anniversary ed. Westport, CT: Praeger, 2003.

Crutzen, Paul, and Eugene Stoermer. "The 'Anthropocene.'" *Global Change Newsletter* 41 (May 2000): 17–18.

Curtis, Kent A. *Gambling on Ore: The Nature of Metal Mining in the United States, 1860–1910*. Boulder: University Press of Colorado, 2013.

Cushman, Gregory T. *Guano and the Opening of the Pacific World: A Global Ecological History*. New York: Cambridge University Press, 2013.

Davis, Georgiann. *Contesting Intersex: The Dubious Diagnosis*. New York: New York University Press, 2015.

Davis, Mike. *Ecology of Fear: Los Angeles and the Imagination of Disaster*. New York: Metropolitan Books, 1998.

Deutsch, Tracey. *Building a Housewife's Paradise: Gender, Politics, and American Grocery Stores in the Twentieth Century*. Chapel Hill: University of North Carolina Press, 2010.

Diamond, Jared. *Guns, Germs, and Steel: The Fates of Human Societies*. New York: W. W. Norton, 1997.

Dickens, Charles. *Hard Times*. London: Electric Book, 2001.

Disco, Nil, and Eda Kranakis, eds. *Cosmopolitan Commons: Sharing Resources and Risks across Borders*. Cambridge, MA: MIT Press, 2013.

Dolin, Eric. *Leviathan: The History of Whaling in America*. New York: W. W. Norton, 2007.

Douglas, Mary. *Purity and Danger: An Analysis of Concepts of Pollution and Taboo*. New York: Praeger, 1966.

Dowie, Mark. *Conservation Refugees: The Hundred-Year Conflict between Global Conservation and Native Peoples*. Cambridge, MA: MIT Press, 2009.

Downs, Laura Lee. *Childhood in the Promised Land: Working-Class Movements and the Colonies de Vacances in France, 1880–1960*. Durham, NC: Duke University Press, 2002.

Dunaway, Finis. *Natural Visions: The Power of Images in American Environmental Reform*. Chicago: University of Chicago Press, 2005.

Dunaway, Finis. *Seeing Green: The Use and Abuse of American Environmental Images*. Chicago: University of Chicago Press, 2015.

Dyl, Joanna. "Lessons from History: Coastal Cities and Natural Disaster." *Management of Environmental Quality: An International Journal* 20, no. 4 (2009): 460–73.

Edelson, S. Max. *Plantation Enterprise in Colonial South Carolina.* Cambridge, MA: Harvard University Press, 2006.

Edgerton, David. *The Shock of the Old: Technology and Global History since 1900.* New York: Oxford University Press, 2011.

Eisenman, T. S. "Frederick Law Olmsted, Green Infrastructure, and the Evolving City." *Journal of Planning History* 12, no. 4 (2013): 287–311.

Elkind, Sarah S. *Bay Cities and Water Politics: The Battle for Resources in Boston and Oakland.* Lawrence: University Press of Kansas, 1998.

Elmore, Bartow J. *Citizen Coke: The Making of Coca-Cola Capitalism.* New York: W. W. Norton, 2014.

Engels, Friedrich. *The Condition of the Working Class in England.* New York: Oxford University Press, 1999.

Ensmenger, Nathan. "The Environmental History of Computing." *Technology and Culture* 59, no. 4 (Supplement, 2018): S7–S33.

Evelyn, John. *Fumifugium; or, The inconveniencie of the aer and smoak of London dissipated together with some remedies humbly proposed by J.E. esq. to His Sacred Majestie, and to the Parliament now assembled.* 1661. Reprinted, Exeter, UK: The Rota at the University of Exeter, 1976.

Everleigh, David J. *Privies and Waterclosets.* London: Shire Library, 2008.

Falck, Zachary J. S. *Weeds: An Environmental History of Urban America.* Pittsburgh: University of Pittsburgh Press, 2010.

Fiege, Mark. *Irrigated Eden: The Making of an Agricultural Landscape in the American West.* Seattle: University of Washington Press, 1999.

Fiege, Mark. "The Weedy West: Mobile Nature, Boundaries, and Common Space in the Montana Landscape." *Western Historical Quarterly* 36, no. 1 (2005): 22–47.

Finger, Thomas D. "Trading Spaces: Transferring Energy and Organizing Power in the Nineteenth-Century Atlantic Grain Trade." In *New Natures: Joining Environmental History with Science and Technology Studies*, edited by Dolly Jørgensen, Finn Arne Jørgensen, and Sara Pritchard, 151–63. Pittsburgh: University of Pittsburgh Press, 2013.

Fitzgerald, Deborah. *The Business of Breeding: Hybrid Corn in Illinois, 1890–1940*. Ithaca, NY: Cornell University Press, 1990.

Fitzgerald, Deborah. *Every Farm a Factory: The Industrial Ideal in American Agriculture*. New Haven, CT: Yale University Press, 2003.

Fortun, Kim. *Advocacy after Bhopal: Environmentalism, Disaster, New Global Orders*. Chicago: University of Chicago Press, 2001.

Fortun, Kim, and Scott Frickel. "Making a Case for Disaster Science and Technology Studies." *An STS Forum on the East Japan Disaster*. Accessed May 15, 2018. https://fukushimaforum.wordpress.com /online-forum-2/online-forum/making-a-case-for-disaster-science -and-technology-studies/.

Foucault, Michel. *Society Must Be Defended: Lectures at the Collège de France, 1975–1976*. New York: St. Martin's Press, 1997.

Frank, Allison Fleig. "The Air Cure Town: Commodifying Mountain Air in Alpine Central Europe." *Central European History* 45, no. 2 (2012): 185–207.

Frank, Alison Fleig. *Oil Empire: Visions of Prosperity in Austrian Galicia*. Cambridge, MA: Harvard University Press, 2005.

Freidberg, Susanne. *French Beans and Food Scares: Culture and Commerce in an Anxious Age*. New York: Oxford University Press, 2004.

Freidberg, Susanne. *Fresh: A Perishable History*. Cambridge, MA: Belknap Press of Harvard University Press, 2009.

Frickel, Scott. *Chemical Consequences: Environmental Mutagens, Scientist Activism, and the Rise of Genetic Toxicology*. New Brunswick, NJ: Rutgers University Press, 2004.

Frickel, Scott, and M. Bess Vincent. "Hurricane Katrina, Contamination, and the Unintended Organization of Ignorance." *Technology in Society* 29, no. 2 (April 2007): 181–88.

Friedel, Robert. "American Bottles: The Road to No Return." *Environmental History* 19, no. 3 (July 2014): 505–27.

Furniss, Philip Jamie. *Metaphors of Waste: Several Ways of Seeing "Development" and Cairo's Garbage Collectors*. New York: Oxford University Press, 2012.

Gabrys, Jennifer. *Digital Rubbish: A Natural History of Electronics*. Ann Arbor: University of Michigan Press, 2011.

Galison, Peter, and Robb Moss, dirs. *Containment*. Film. Redacted Pictures, 2015.

Gammage, Bill. *The Biggest Estate on Earth: How Aborigines Made Australia*. Sydney: Allen & Unwin, 2011.

Gaskell, Elizabeth. *North and South*. London: Chapman & Hall, 1855. Reprinted, London: Penguin, 2011.

Geddes, Patrick. *Cities in Evolution*. London: Williams & Norgate, 1915.

George, Timothy S. "Fukushima in Light of Minamata." *Asia-Pacific Journal* 10 (March 5, 2012). http://apjjf.org/2012/10/11/Timothy-S .-George/3715/article.html.

George, Timothy S. *Minamata: Pollution and the Struggle for Democracy in Postwar Japan*. Cambridge, MA: Harvard University Asia Center, 2001.

Gerber, Michele Stenehjem. *On the Home Front: The Cold War Legacy of the Hanford Nuclear Site*. Lincoln: University of Nebraska Press, 2002.

Geurts, Kathryn Linn. *Culture and the Senses: Bodily Ways of Knowing in an African Community*. Berkeley: University of California Press, 2002.

Gilbert, Erik, and Jonathan Reynolds. *Trading Tastes: Commodity and Cultural Exchange to 1750*. Upper Saddle River, NJ: Prentice Hall, 2006.

Gille, Zsuzsa. *From the Cult of Waste to the Trash Heap of History: The Politics of Waste in Socialist and Postsocialist Hungary*. Bloomington: Indiana University Press, 2007.

Gold, Daniel B., and Judith Helfand, dirs. *Blue Vinyl*. Oley, PA: Bullfrog Films, 2002.

Gorman, Hugh S. *Redefining Efficiency: Pollution Concerns, Regulatory Mechanisms, and Technological Change in the U.S. Petroleum Industry*. Akron, OH: University of Akron Press, 2001.

Gorman, Hugh S. *The Story of N: A Social History of the Nitrogen Cycle and the Challenge of Sustainability*. New Brunswick, NJ: Rutgers University Press, 2013.

Gorman, Hugh S., and Betsy Mendelsohn. "Where Does Nature End and Culture Begin? Converging Themes in the History of Technology and Environmental History." In *The Illusory Boundary: Environment and Technology in History*, edited by Martin Reuss and Stephen H. Cutcliffe, 265–90. Charlottesville: University of Virginia Press, 2010.

Gowing, Laura. *Women, Touch, and Power in Seventeenth-Century England*. New Haven, CT: Yale University Press, 2003.

Greene, Ann Norton. *Horses at Work: Harnessing Power in Industrial America*. Cambridge, MA: Harvard University Press, 2008.

Gregson, Nicky, and Mike Crang. "From Waste to Resource: The Trade in Wastes and Global Recycling Economies." *Annual Review of Environment and Resources* 40, no. 1 (2015): 151–76.

Grossman, Elizabeth. *High Tech Trash: Digital Devices, Hidden Toxics, and Human Health*. Washington, DC: Island Press, 2006.

Grosz, Elizabeth. "Bodies and Knowledges: Feminism and the Crisis of Reason." In *Space, Time, and Perversion*, 25–44. London: Routledge, 1995.

Gunderson, Lance H., and C. S. Hollings, eds. *Panarchy: Understanding Transformations in Human and Natural Systems*. Washington, DC: Island Press, 2002.

Haave, Marte, Erik Ropstad, Andrew E. Derocher, Elisabeth Lie, Ellen Dahl, Oystein Wiig, Janneche U. Skaare, and Bjorn Munro Jenssen. "Polychlorinated Biphenyls and Reproductive Hormones in Female Polar Bears at Svalbard." *Environmental Health Perspectives* 111, no. 4 (2003): 431–36.

Halberstam, Jack. *Trans\*: A Quick and Quirky Account of Gender Variability*. Berkeley: University of California Press, 2018.

Hamblin, Jacob Darwin. *Poison in the Well: Radioactive Waste in the Oceans at the Dawn of the Nuclear Age*. New Brunswick, NJ: Rutgers University Press, 2008.

Hamilton, Shane. *Trucking Country: The Road to America's Wal-Mart Economy*. Princeton, NJ: Princeton University Press, 2014.

Hamlin, Christopher. "Edwin Chadwick and the Engineers, 1842–1854: Systems and Antisystems in the Pipe-and-Brick Sewers War." *Technology and Culture* 33, no. 4 (1992): 680–709.

Hannickel, Erica. *Empire of Vines: Wine Culture in America*. Philadelphia: University of Pennsylvania Press, 2013.

Haraway, Donna J. "Anthropocene, Capitalocene, Plantationocene, Chthulucene: Making Kin." *Environmental Humanities* 6 (2015): 159–65.

Haraway, Donna J. *The Companion Species Manifesto: Dogs, People, and Significant Otherness*. Chicago: Prickly Paradigm Press, 2003.

Haraway, Donna J. "Situated Knowledges: The Science Question in Feminism and the Privilege of Partial Perspectives." *Feminist Studies* 14, no. 3 (1988): 575–99.

Haraway, Donna J. *Staying with the Trouble: Making Kin in the Chthulucene*. Durham, NC: Duke University Press, 2016.

Haraway, Donna J. *When Species Meet*. Minneapolis: University of Minnesota Press, 2008.

Harlan, Jack R. *Crops and Man*. Madison, WI: American Society of Agronomy, 1975.

Harrison, Jill Lindsey. *Pesticide Drift and the Pursuit of Environmental Justice*. Cambridge, MA: MIT Press, 2011.

Hayles, N. Katherine. *How We Became Posthuman: Virtual Bodies in Cybernetics, Literature, and Informatics*. Chicago: University of Chicago Press, 1999.

Hays, Samuel P. "Science and Values: Lead in Historical Perspective." In

*Explorations in Environmental History*, 291–311. Pittsburgh: University of Pittsburgh Press, 1998.

Hecht, Gabrielle. "The African Anthropocene." *Aeon*, February 6, 2018. https://aeon.co/essays/if-we-talk-about-hurting-our-planet-who -exactly-is-the-we.

Hecht, Gabrielle. *Being Nuclear: Africans and the Global Uranium Trade*. Cambridge, MA: MIT Press, 2012.

Hecht, Gabrielle. "Globalization meets Frankenstein? Reflections on Terrorism and Technopolitics in the Nuclear Age." *History and Technology* 19, no. 1 (2003): 1–8.

Hecht, Gabrielle. "Interscalar Vehicles for an African Anthropocene: On Waste, Temporality, and Violence." *Cultural Anthropology* 33, no. 1 (2018): 109–41.

Hecht, Gabrielle. "Invisible Production and the Production of Invisibility: Cleaning, Maintenance, and Mining in the Nuclear Sector." In *Routledge Handbook of Science, Technology, and Society*, edited by Daniel Kleinman and Kelly Moore, 346–61. Abingdon, UK: Routledge, 2014.

Hecht, Gabrielle. "Nuclear Nomads: A Look at the Subcontracted Heroes." *Bulletin of the Atomic Scientists* (January 9, 2012). http://thebulletin.org/nuclear-nomads-look-subcontracted-heroes.

Hecht, Gabrielle. *The Radiance of France: Nuclear Power and National Identity after World War II*. Cambridge, MA: MIT Press, 1998.

Hecht, Gabrielle. "The Work of Invisibility: Radiation Hazards and Occupational Health in South African Uranium Production." *International Labor and Working-Class History* 81 (spring 2012): 94–113.

Helmreich, Stefan. "Underwater Music: Tuning Composition to the Sounds of Science." In *The Oxford Handbook of Sound Studies*, edited by Trevor Pinch and Karin Bijsterveld, 151–75. New York: Oxford University Press, 2012.

Hempton, Gordon, and John Grossman. *One Square Inch of Silence:*

*One Man's Search for Natural Silence in a Noisy World*. New York: Free Press, 2009.

Hill, Sarah. "Garbage Here, Recycling There: Used-Up Materials across North American Borders." In *Actions: Playing, Gardening, Recycling and Walking*, edited by Giovanna Borasi and Mirko Zardini, 154–63. Montreal: Canadian Center for Architecture, 2006.

Hindle, Brooke, and Steven Lubar. *Engines of Change: The American Industrial Revolution, 1790–1860*. Washington, DC: Smithsonian Institution Press, 1986.

Hindmarsh, Richard, ed. *Nuclear Disaster at Fukushima Daiichi: Social, Political and Environmental Issues*. New York: Routledge, 2013.

Hogan, David Gerard. *Selling 'Em by the Sack: White Castle and the Creation of American Food*. New York: New York University Press, 1997.

Hogan, Michael J., ed. *Hiroshima in History and Memory*. Cambridge: Cambridge University Press, 1996.

Höhler, Sabine. *Spaceship Earth in the Environmental Age, 1960–1990*. London: Pickering & Chatto, 2015.

Honeyman, Katrina. *Women, Gender, and Industrialization in England, 1700–1870*. New York: St. Martin's Press, 2000.

Horn, Jeff. *The Path Not Taken: French Industrialization in the Age of Revolution, 1750–1830*. Cambridge, MA: MIT Press, 2006.

Horowitz, Roger. "Making the Chicken of Tomorrow: Reworking Poultry as Commodities and as Creatures, 1945–1990." In *Industrializing Organisms: Introducing Evolutionary History*, edited by Susan Schrepfer and Philip Scranton, 215–36. New York: Routledge, 2004.

Howe, Joshua P. *Behind the Curve: Science and the Politics of Global Warming*. Seattle: University of Washington Press, 2014.

Howes, David. *Empire of the Senses: The Sensual Culture Reader*. New York: Berg, 2005.

Hoy, Suellen. *Chasing Dirt: The American Pursuit of Cleanliness*. New York: Oxford University Press, 1995.

Huber, Matthew. *Lifeblood: Oil, Freedom, and the Forces of Capital.* Minneapolis: University of Minnesota Press, 2013.

Hughes, Thomas Parke. *Human-Built World: How to Think about Technology and Culture.* Chicago: University of Chicago Press, 2004.

Hughes, Thomas Parke. *Networks of Power: Electrification in Western Society, 1880–1930.* Baltimore: Johns Hopkins University Press, 1983.

Hurley, Andrew. *Environmental Inequalities: Class, Race, and Industrial Pollution in Gary, Indiana, 1945–1980.* Chapel Hill: University of North Carolina Press, 1995.

Hutchings, Robert. "Consuming Nature: Fresh Fruit, Processed Juice, and the Re-making of the Florida Orange, 1877–2014." PhD dissertation, Carnegie Mellon University, 2014.

Igler, David. *Industrial Cowboys: Miller & Lux and the Transformation of the Far West, 1850–1920.* Berkeley: University of California Press, 2001.

Isenberg, Andrew. *Destruction of the Bison: An Environmental History, 1750–1920.* New York: Cambridge University Press, 2000.

Jack, Caroline, and Stephanie Steinhardt. "Atomic Anxiety and the Tooth Fairy: Citizen Science in the Midcentury Midwest." *Appendix* 2, no. 4 (October 2014). http://theappendix.net/issues/2014/10/atomic-anxiety-and-the-tooth-fairy-citizen-science-in-the-mid century-midwest.

Jackson, Jeffrey H. *Paris Under Water: How the City of Light Survived the Great Flood of 1910.* New York: Palgrave Macmillan, 2010.

Jackson, Mark. *Allergy: The History of a Modern Malady.* London: Reaktion, 2006.

Jackson, Mark. *Asthma: The Biography.* Oxford: Oxford University Press, 2009.

Jacobs, Jane. *The Death and Life of Great American Cities.* New York: Vintage, 1961.

Jacoby, Karl. *Crimes Against Nature: Squatters, Poachers, Thieves, and the Hidden History of American Conservation.* Berkeley: University of California Press, 2003.

Jakle, John A. *City Lights: Illuminating the American Night*. Baltimore: Johns Hopkins University Press, 2001.

Jarzabkowski, Paula, Rebecca Bednarek, and Paul Spee. *Making a Market for Acts of God: The Practice of Risk-Trading in the Global Reinsurance Industry*. Oxford: Oxford University Press, 2015.

Jasanoff, Sheila. "Bhopal's Trials of Knowledge and Ignorance." *Isis* 98, no. 2 (2007): 344–50.

Jasanoff, Sheila, ed. *Learning from Disaster: Risk Management after Bhopal*. Philadelphia: University of Pennsylvania Press, 1994.

Jones, Christopher. *Routes of Power: Energy and Modern America*. Cambridge, MA: Harvard University Press, 2014.

Jones, Nicola. "How the World Passed a Carbon Threshold and Why It Matters." *YaleEnvironment360*, January 26, 2017. https://e360.yale .edu/features/how-the-world-passed-a-carbon-threshold-400ppm -and-why-it-matters.

Jørgensen, Dolly. "Cooperative Sanitation: Managing Streets and Gutters in Late Medieval England and Scandinavia." *Technology and Culture* 49, no. 3 (2008): 547–67.

Jørgensen, Dolly. "Local Government Responses to Urban River Pollution in Late Medieval England." *Water History* 2, no. 1 (2010): 35–52.

Jørgensen, Dolly. "The Medieval Sense of Smell, Stench, and Sanitation." In *Les cinq sens de la ville du Moyen Âge à nos jours*, edited by Ulrike Krampl, Robert Beck, and Emmanuelle Retaillaud-Bajac, 301–13. Tours: Presses Universitaires François-Rabelais, 2013.

Jørgensen, Dolly. "Modernity and Medieval Muck." *Nature and Culture* 9, no. 3 (2014): 225–37.

Jørgensen, Dolly. "What to Do with Waste? The Challenges of Waste Disposal in Two Late Medieval Towns." In *Living Cities: An Anthology in Urban Environmental History*, edited by Matthias Legnér and Sven Lilja, 34–55. Stockholm: Forskningsrådet Formas, 2010.

Jørgensen, Finn Arne. "The Armchair Traveler's Guide to Digital Environmental Humanities." *Environmental Humanities* 4, no. 1 (2014): 95–112.

Jørgensen, Finn Arne. *Making a Green Machine: The Infrastructure of Beverage Container Recycling*. New Brunswick, NJ: Rutgers University Press, 2011.

Jørgensen, Finn Arne. "Walking with GPS." In *Methodological Challenges in Nature-Culture and Environmental History Research*, edited by Jocelyn Thorpe, Stephanie Rutherford, and L. Anders Sandberg, 284–97. New York: Routledge, 2016.

Josephson, Paul R. *Industrialized Nature: Brute Force Technology and the Transformation of the Natural World*. Washington, DC: Island Press, 2002.

Josephson, Paul R. *Motorized Obsessions: Life, Liberty, and the Small-Bore Engine*. Baltimore: Johns Hopkins University Press, 2007.

Josephson, Paul R. "The Ocean's Hot Dog: The Development of the Fish Stick." *Technology and Culture* 49, no. 1 (2008): 41–61.

Kahn, Peter H., Jr. "Children's Affiliations with Nature: Structure, Development, and the Problem of Environmental Generational Amnesia." In *Children and Nature: Psychological, Sociocultural, and Evolutionary Investigations*, edited by Peter H. Kahn Jr. and Stephen R. Keller, 93–116. Cambridge, MA: MIT Press, 2002.

Kaijser, Arne. "Under a Common Acid Sky: Negotiating Transboundary Air Pollution in Europe." In *Cosmopolitan Commons: Sharing Resources and Risks across Borders*, edited by Nil Disco and Eda Kranakis, 213–42. Cambridge, MA: MIT Press, 2013.

Keirns, Carla. "Short of Breath: A Social and Intellectual History of Asthma in the United States." PhD dissertation, University of Pennsylvania, 2004.

Keller, Richard C. *Fatal Isolation: The Devastating Paris Heat Wave of 2003*. Chicago: University of Chicago Press, 2015.

Kelman, Ari. "Boundary Issues: Clarifying New Orleans's Murky Edges." *Journal of American History* 94, no. 3 (2007): 695–703.

Kelman, Ari. "Nature Bats Last: Some Recent Works on Technology and Urban Disaster." *Technology and Culture* 47, no. 2 (2006): 391–402.

Kelman, Ari. *A River and Its City: The Nature of Landscape in New*

*Orleans*. Berkeley: University of California Press, 2003. Reprinted, with a new preface by the author, 2006.

Kingsolver, Barbara. *Animal, Vegetable, Miracle: A Year of Food Life*. New York: HarperCollins, 2007.

Kirksey, Eben, ed. *The Multispecies Salon*. Durham, NC: Duke University Press, 2014.

Kirksey, Eben, and Stefan Helmreich. "The Emergence of Multispecies Ethnography." *Cultural Anthropology* (June 14, 2010). https:// culanth.org/fieldsights/277-the-emergence-of-multispecies -ethnography.

Klein, Naomi. *The Shock Doctrine: The Rise of Disaster Capitalism*. New York: Metropolitan Books, 2007.

Kline, Ronald. "Construing 'Technology' as 'Applied Science': Public Rhetoric of Scientists and Engineers in the United States, 1880– 1945." *Isis* 86, no. 2 (1995): 194–221.

Kline, Ronald, and Trevor Pinch. "Users as Agents of Technological Change: The Social Construction of the Automobile in the Rural United States." *Technology and Culture* 37, no. 4 (1996): 763–95.

Kloppenburg, Jack Ralph, Jr. *First the Seed: The Political Economy of Plant Biotechnology, 1492–2000*. New York: Cambridge University Press, 1988.

Knowles, Scott Gabriel. "Learning from Disaster? The History of Technology and the Future of Disaster Research." *Technology and Culture* 55, no. 4 (2014): 773–84.

Kohler, Robert E. *Lords of the Fly: Drosophila Genetics and the Experimental Life*. Chicago: University of Chicago Press, 1994.

Krech, Shepard, III. *The Ecological Indian: Myth and History*. New York: W. W. Norton, 1999.

Kuriyama, Shigehisa. *The Expressiveness of the Body and the Divergence of Greek and Chinese Medicine*. New York: Zone Books, 1999.

Kuriyama, Shigehisa, ed. *The Imagination of the Body and the History of Bodily Experience*. Kyoto: International Research Center for Japanese Studies, 2001.

Kurlansky, Mark. *Salt: A World History.* New York: Walker, 2002.

Kyba, Christopher C. M., Thomas Ruhtz, Jürgen Fischer, and Franz Hölker. "Red Is the New Black: How the Color of Urban Skyglow Varies with Cloud Cover." *Monthly Notices of the Royal Astronomical Society* 425, no. 1 (2012): 701–8.

Lacquer, Thomas Walter. *Making Sex: Body and Gender from the Greeks to Freud.* Cambridge, MA: Harvard University Press, 1990.

Laemmli, Whitney E. "A Case in Pointe: Romance and Regimentation at the New York City Ballet." *Technology and Culture* 56, no. 1 (2015): 1–27.

Lakwete, Angela. *Inventing the Cotton Gin: Machine and Myth in Antebellum America.* Baltimore: Johns Hopkins University Press, 2003.

Langston, Nancy. "The Retreat from Precaution: Regulating Diethylstilbestrol (DES), Endocrine Disruptors, and Environmental Health." *Environmental History* 13, no. 1 (2008): 41–65.

Langston, Nancy. *Toxic Bodies: Hormone Disruptors and the Legacy of DES.* New Haven, CT: Yale University Press, 2010.

Latour, Bruno. *Reassembling the Social: An Introduction to Actor-Network-Theory.* New York: Oxford University Press, 2005.

Latour, Bruno. *Science in Action: How to Follow Scientists and Engineers through Society.* Cambridge, MA: Harvard University Press, 1987.

Latour, Bruno. *We Have Never Been Modern.* Translated by Catherine Porter. Cambridge, MA: Harvard University Press, 1993.

LeCain, Timothy J. *Mass Destruction: The Men and Giant Mines That Wired America and Scarred the Planet.* New Brunswick, NJ: Rutgers University Press, 2009.

Leech, Brian. *The City That Ate Itself: Butte, Montana, and Its Expanding Berkeley Pit.* Reno: University of Nevada Press, 2018.

Lerman, Nina E., Ruth Oldenziel, and Arwen P. Mohun, eds. *Gender and Technology: A Reader.* Baltimore, MD: Johns Hopkins University Press, 2003.

Li, Lillian M. *Fighting Famine in North China: State, Market, and Environmental Decline, 1690s–1990s.* Stanford, CA: Stanford University Press, 2007.

Liboiron, Max. "Redefining Pollution and Action: The Matter of Plastics." *Journal of Material Culture* 21, no. 1 (2016): 87–110.

Liboiron, Max. "Redefining Pollution: Plastics in the Wild." PhD dissertation, New York University, 2012.

Licht, Walter. *Industrializing America: The Nineteenth Century.* Baltimore: Johns Hopkins University Press, 1995.

Loewenstein, Antony. *Disaster Capitalism: Making a Killing Out of Catastrophe.* New York: Verso, 2015.

Longhurst, James. *Citizen Environmentalists.* Lebanon, NH: University Press of New England, 2010.

Lowood, Henry E. "The Calculating Forester: Quantification, Cameral Science, and the Emergence of Scientific Forestry Management in Germany." In *The Quantifying Spirit in the 18th Century,* edited by Tore Frängsmyr, J. L. Heilbron, and Robin E. Rider, 315–42. Berkeley: University of California Press, 1990.

Lucsko, David N. *Junkyards, Gearheads, and Rust: Salvaging the Automotive Past.* Baltimore: Johns Hopkins University Press, 2016.

MacBride, Samantha. *Recycling Reconsidered: The Present Failure and Future Promise of Environmental Action in the United States.* Cambridge, MA: MIT Press, 2011.

MacFarlane, Alison M., and Rodney C. Ewing, eds. *Uncertainty Underground: Yucca Mountain and the Nation's High-Level Nuclear Waste.* Cambridge, MA: MIT Press, 2006.

MacLachlan, Ian. *Kill and Chill: Restructuring Canada's Beef Commodity Chain.* Toronto: University of Toronto Press, 2001.

Maher, Neil M. "Body Counts: Tracking the Human Body through Environmental History." In *A Companion to American Environmental History,* edited by Douglas Cazaux Sackman, 163–80. Malden, MA: Wiley-Blackwell, 2010.

Maher, Neil M. *Nature's New Deal: The Civilian Conservation Corps* and the Origins of the American Environmental Movement. New York: Oxford University Press, 2009.

Maher, Neil M. "Shooting the Moon: How NASA Earth Photographs Changed the World." *Environmental History* 9, no. 3 (2004): 526–31.

Maines, Rachel P. *Asbestos and Fire: Technological Trade-Offs and the Body at Risk*. New Brunswick, NJ: Rutgers University Press, 2005.

Maines, Rachel P. *The Technology of Orgasm: "Hysteria," the Vibrator, and Women's Sexual Satisfaction*. Baltimore, MD: Johns Hopkins University Press, 1999.

Marks, Robert. *The Origins of the Modern World: A Global and Ecological Narrative*. Lanham, MD: Rowman & Littlefield, 2002.

Marran, Christine L. "Contamination: From Minamata to Fukushima." *Asia-Pacific Journal* 9 (May 9, 2011). http://apjjf.org/2011/9/19/Christine-Marran/3526/article.html.

Martín, Carlos E. "Mechanization and 'Mexicanization': Racializing California's Agricultural Technology." *Science as Culture* 10, no. 3 (2001): 301–26.

Martin, Laura J. "Proving Grounds: Ecological Fieldwork in the Pacific and the Materialization of Ecosystems." *Environmental History* 23, no. 3 (2018): 567–92.

Martineau, Belinda. *First Fruit: The Creation of the Flavr Savr™ Tomato and the Birth of Biotech Food*. New York: McGraw Hill, 2001.

Marx, Karl. *Capital*. Chicago: Encyclopaedia Britannica, 1952.

Marx, Leo. "The Idea of 'Technology' and Postmodern Pessimism." In *Does Technology Drive History? The Dilemma of Technological Determinism*, edited by Merritt Roe Smith and Leo Marx, 237–57. Cambridge, MA: MIT Press, 1994.

Marx, Leo. "Technology: The Emergence of a Hazardous Concept." *Technology and Culture* 51, no. 3 (2010): 561–77.

Matchar, Emily. "Is Michael Pollan a Sexist Pig?" *Salon*, April 28, 2013. https://www.salon.com/2013/04/28/is_michael_pollan_a_sexist_pig/.

Mauch, Christof, and Thomas Zeller, eds. *Rivers in History: Perspectives on Waterways in Europe and North America.* Pittsburgh: University of Pittsburgh Press, 2008.

Mavhunga, Clapperton Chakanestsa. *Transient Workspaces: Technologies of Everyday Innovation in Zimbabwe.* Cambridge, MA: MIT Press, 2014.

McCann, James C. *Maize and Grace: Africa's Encounter with a New World Crop, 1500–2000.* Cambridge, MA: Harvard University Press, 2005.

McDonough, William, and Michael Braungart. *Cradle to Cradle: Remaking the Way We Make Things.* New York: North Point Press, 2002.

McDonough, William, and Michael Braungart. *The Upcycle: Beyond Sustainability—Designing for Abundance.* New York: North Point Press, 2013.

McEvoy, Arthur F. "Working Environments: An Ecological Approach to Industrial Health and Safety." *Technology and Culture* 36, no. 2 (1995): 145–72.

McGaw, Judith. *Most Wonderful Machine: Mechanization and Change in Berkshire Paper Making, 1801–1885.* Princeton, NJ: Princeton University Press, 1987.

McGurty, Eileen. *Transforming Environmentalism: Warren County, PCBs, and the Origins of Environmental Justice.* New Brunswick, NJ: Rutgers University Press, 2007.

McNeill, John R. *Something New under the Sun: An Environmental History of the Twentieth-Century World.* New York: W. W. Norton, 2000.

McNeill, John R., and William H. McNeill. *The Human Web: A Bird's-Eye View of World History.* New York: W. W. Norton, 2003.

McPhee, John. *The Control of Nature.* New York: Farrar, Straus & Giroux, 1989.

McShane, Clay. *Down the Asphalt Path: The Automobile and the American City.* New York: Columbia University Press, 1995.

McShane, Clay, and Joel A. Tarr. *The Horse in the City: Living Machines in the Nineteenth Century*. Baltimore: Johns Hopkins University Press, 2007.

Medina, Martin. *The World's Scavengers: Salvaging for Sustainable Consumption and Production*. Lanham, MD: AltaMira, 2007.

Melillo, Edward D. "The First Green Revolution: Debt Peonage and the Making of the Nitrogen Fertilizer Trade, 1840–1930." *American Historical Review* 117, no. 4 (2012): 1028–60.

Melosi, Martin V. *Garbage in the Cities: Refuse, Reform, and the Environment*. Revised ed. Pittsburgh: University of Pittsburgh Press, 2004.

Melosi, Martin V. *The Sanitary City: Environmental Services in Urban America from Colonial Times to Present*. Pittsburgh: University of Pittsburgh Press, 2008.

Merchant, Carolyn. *Ecological Revolutions: Nature, Gender, and Science in New England*. Chapel Hill: University of North Carolina Press, 2010.

Minter, Adam. *Junkyard Planet: Travels in the Billion-Dollar Trash Trade*. New York: Bloomsbury, 2013.

Mintz, Sidney. *Sweetness and Power: The Place of Sugar in Modern History*. New York: Penguin, 1986.

Mitchell, Timothy. *Carbon Democracy: Political Power in the Age of Oil*. New York: Verso, 2011.

Mitchell, Timothy. *Rule of Experts: Egypt, Techno-politics, Modernity*. Oakland: University of California Press, 2002.

Mitman, Gregg. *Breathing Space: How Allergies Shape Our Lives and Landscapes*. New Haven, CT: Yale University Press, 2007.

Mitman, Gregg, Michelle Murphy, and Christopher Sellers. "Introduction: A Cloud over History." *Osiris* 19, no. 1 (2004): 1–17.

Molotch, Harvey L., and Laura Norén, eds. *Toilet: Public Restrooms and the Politics of Sharing*. New York: New York University Press, 2010.

Moon, David. *The Plough That Broke the Steppes: Agriculture and Environment on Russia's Grasslands, 1700–1914*. New York: Oxford University Press, 2013.

Moore, Jason W. *Capitalism in the Web of Life: Ecology and the Accumulation of Capital*. New York: Verso, 2015.

Moore, Jason W. "The Capitalocene, Part I: On the Nature and Origins of Our Ecological Crisis." *Journal of Peasant Studies* 44, no. 3 (2017): 594–630.

Moore, Jason W. "The Capitalocene, Part II: Accumulation by Appropriation and the Centrality of Unpaid Work/Energy." *Journal of Peasant Studies* 45, no. 2 (2018): 237–79.

Moore, Jason W., ed. *Anthropocene or Capitalocene? Nature, History, and the Crisis of Capitalism*. Oakland, CA: PM Press, 2016.

Mukerji, Chandra. *Impossible Engineering: Technology and Territoriality on the Canal du Midi*. Princeton, NJ: Princeton University Press, 2009.

Murphy, Michelle. *Sick Building Syndrome and the Problem of Uncertainty: Environmental Politics, Technoscience, and Women Workers*. Durham, NC: Duke University Press, 2006.

Muscolino, Micah. *Fishing Wars and Environmental Change in Late Imperial and Modern China*. Cambridge, MA: Harvard University Press, 2009.

Nagle, Robin. *Picking Up: On the Streets and Behind the Trucks with the Sanitation Workers of New York City*. New York: Farrar, Straus & Giroux, 2013.

Nash, Linda. "The Agency of Nature or the Nature of Agency?" *Environmental History* 10, no. 1 (2005): 67–69.

Nash, Linda. "The Fruits of Ill-Health: Pesticides and Workers' Bodies in Post–World War II California." *Osiris* 19 (2004): 203–19.

Nash, Linda. *Inescapable Ecologies: A History of Environment, Disease, and Knowledge*. Berkeley: University of California Press, 2006.

Neumann, Roderick P. *Imposing Wilderness: Struggles over Livelihood and Nature Preservation in Africa*. Berkeley: University of California Press, 2002.

Newman, Richard. "Darker Shades of Green: Love Canal, Toxic Autobiography, and American Environmental Writing." In *Histories of the*

*Dustheap: Waste, Material Cultures, Social Justice*, edited by
Stephanie Foote and Elizabeth Mazzolini, 21–48. Cambridge, MA:
MIT Press, 2012.

Nixon, Rob. *Slow Violence and the Environmentalism of the Poor.*
Cambridge, MA: Harvard University Press, 2011.

Nye, David E. *Electrifying America: Social Meanings of a New Technology, 1880–1940.* Cambridge, MA: MIT Press, 1990.

Nye, David E. "The Transformation of American Urban Space: Early
Electric Lighting, 1875–1915." In *Urban Lighting, Light Pollution,
and Society*, edited by Josiane Meier, Ute Hasenohrl, Katharina
Krause, and Merle Pottharst, 30–45. New York: Routledge, 2015.

Nye, David E. *When the Lights Went Out: A History of Blackouts in
America.* Cambridge, MA: MIT Press, 2010.

Nystrom, Eric C. *Seeing Underground: Maps, Models, and Mining
Engineering in America.* Reno: University of Nevada Press, 2014.

Ogle, Maureen. *All the Modern Conveniences: American Household
Plumbing, 1840–1890.* Baltimore: Johns Hopkins University Press,
2000.

Oleson, John Peter, ed. *Oxford Handbook of Engineering and Technology in the Classical World.* New York: Oxford University Press, 2008.

Olson, Valerie. "NEOSpace: The Solar System's Emerging Environmen-
tal History and Politics." In *New Natures: Joining Environmental
History with Science and Technology Studies*, edited by Dolly
Jørgensen, Finn Arne Jørgensen, and Sara B. Pritchard, 195–211.
Pittsburgh: Pittsburgh University Press, 2013.

Olson, Valerie A. "Political Ecology in the Extreme: Asteroid Activism
and the Making of an Environmental Solar System." *Anthropological
Quarterly* 85, no. 4 (2012): 1027–44.

Oreskes, Naomi. "Why I Am a Presentist." *Science in Context* 26, no. 4
(2013): 595–609.

Oreskes, Naomi, and Eric M. Conway. *The Collapse of Western Civilization: A View From the Future.* New York: Columbia University Press,
2014.

Oreskes, Naomi, and Eric M. Conway. *Merchants of Doubt: How a Handful of Scientists Obscured the Truth on Issues from Tobacco Smoke to Global Warming*. New York: Bloomsbury, 2011.

Ott, Cindy. *Pumpkin: The Curious History of an American Icon*. Seattle: University of Washington Press, 2012.

Otter, Chris. *The Victorian Eye: A Political History of Light and Vision in Britain, 1800–1910*. Chicago: University of Chicago Press, 2008.

Otter, Chris, Nicholas Breyfogle, John L. Brooke, Mari K. Webel, Matthew Klingle, Andrew Price-Smith, Brett L. Walker, and Linda Nash. "Forum: Technology, Ecology, and Human Health since 1850." *Environmental History* 20, no. 4 (2015): 710–804.

Pacey, Arnold. *Technology in World Civilization*. Cambridge, MA: MIT Press, 1990.

Pacyga, Dominic A. *Slaughterhouse: Chicago's Union Stock Yard and the World It Made*. Chicago: University of Chicago Press, 2015.

Parr, Joy. "Our Bodies and Our Histories of Technology and the Environment." In *The Illusory Boundary*, edited by Martin Reuss and Stephen H. Cutcliffe, 26–42. Charlottesville: University of Virginia Press, 2010.

Parr, Joy. *Sensing Changes: Technologies, Environments, and the Everyday, 1953–2003*. Vancouver: University of British Columbia Press, 2010.

Parrinello, Giacomo. *Fault Lines: Earthquakes and Urbanism in Modern Italy*. New York: Berghahn Books, 2015.

Pauly, Daniel. "Anecdotes and the Shifting Baseline Syndrome of Fisheries." *Trends in Ecology and Evolution* 10, no. 10 (1995): 430.

Paxson, Heather. *The Life of Cheese: Crafting Food and Value in America*. Berkeley: University of California Press, 2012.

Pearson, Chris. "Between Instinct and Intelligence: Harnessing Police Dog Agency in Early Twentieth-Century Paris." *Comparative Studies in Society and History* 58, no. 2 (2016): 463–90.

Pearson, Chris. *Sniffing the Past—Dogs and History* (blog). https://sniffingthepast.wordpress.com/.

Pellow, David. *Garbage Wars: The Struggle for Environmental Justice in Chicago.* Cambridge, MA: MIT Press, 2002.

Pellow, David. *Resisting Global Toxics: Transnational Movements for Environmental Justice.* Cambridge, MA: MIT Press, 2007.

Pellow, David, and Lisa Sun-Hee Park. *The Silicon Valley of Dreams: Environmental Justice, Immigrant Workers, and the High-Tech Global Economy.* New York: New York University Press, 2002.

Penney, Matthew. "Nuclear Nationalism and Fukushima." *Asia-Pacific Journal* 10, no. 11 (March 12, 2012). https://apjjf.org/2012/10/11/Matthew-Penney/3712/article.html.

Perdue, Peter C. "Perdue on Pomeranz." Review of *The Great Divergence: China, Europe, and the Making of the Modern World Economy,* by Kenneth Pomeranz. *H-World,* August 2000. https://networks.h-net.org/node/20292/reviews/21064/perdue-pomeranz-great-divergence-china-europe-and-making-modern-world.

Perkins, John H. *Geopolitics and the Green Revolution: Wheat, Genes, and the Cold War.* New York: Oxford University Press, 1997.

Perrow, Charles. *The Next Catastrophe: Reducing Our Vulnerabilities to Natural, Industrial, and Terrorist Disasters.* Princeton, NJ: Princeton University Press, 2011.

Perrow, Charles. *Normal Accidents: Living with High-Risk Technologies.* 2nd ed. Princeton, NJ: Princeton University Press, 1999.

Perry, Stuart. *Collecting Garbage: Dirty Work, Clean Jobs, Proud People.* New Brunswick, NJ: Transaction, 1998.

Petrick, Gabriella M. "The Arbiters of Taste: Producers, Consumers, and the Industrialization of Taste in America." PhD dissertation, University of Delaware, 2006.

Petrick, Gabriella M. "The Asian Roots of Umami: The 'Fifth' Taste Central to Thanksgiving Fare." *The Conversation,* November 25, 2016. http://theconversation.com/the-asian-roots-of-umami-the-fifth-taste-central-to-thanksgiving-fare-50699.

Petrick, Gabriella M. "'Like Ribbons of Green and Gold': Industrializing

Lettuce and the Quest for Quality in the Salinas Valley, 1920–1965."
*Agricultural History* 80, no. 3 (2006): 269–95.

Petrick, Gabriella M., and Gerard J. Fitzgerald. "In Good Taste: Rethinking American History with Our Palates." *Journal of American History* 95, no. 2 (2008): 392–404.

Petryna, Adriana. *Life Exposed: Biological Citizens after Chernobyl.* Princeton: Princeton University Press, 2002.

Pfaffenberger, Bryan. "The Harsh Facts of Hydraulics: Technology and Society in Sri Lanka's Colonization Schemes." *Technology and Culture* 31, no. 3 (1990): 361–97.

Pilcher, Jeffrey M. *The Sausage Rebellion: Public Health, Private Enterprise, and Meat in Mexico City, 1890–1917.* Albuquerque: University of New Mexico Press, 2006.

Pinch, Trevor, and Karin Bijsterveld, eds. *The Oxford Handbook of Sound Studies.* New York: Oxford University Press, 2012.

Piper, Liza. "Subterranean Bodies: Mining the Large Lakes of Northwest Canada, 1921–1960." *Environment and History* 13, no. 2 (2007): 155–86.

Pisani, Donald J. *From the Family Farm to Agribusiness: The Irrigation Crusade in California and the West, 1850–1931.* Berkeley: University of California Press, 1984.

Pollan, Michael. *The Omnivore's Dilemma: A Natural History of Four Meals.* New York: Penguin, 2006.

Pollan, Michael. "Some of My Best Friends Are Germs." *New York Times Magazine* (May 15, 2013).

Pomeranz, Kenneth. *The Great Divergence: Europe, China, and the Making of the Modern World Economy.* Princeton, NJ: Princeton University Press, 2000.

Powell, Miles A. *Vanishing America: Species Extinction, Racial Peril, and the Origins of Conservation.* Cambridge, MA: Harvard University Press, 2016.

Pritchard, Sara B. *Confluence: The Nature of Technology and the Remaking of the Rhône.* Cambridge, MA: Harvard University Press, 2011.

Pritchard, Sara B. "Envirotech Methods: Looking Back, Looking Beyond?" Paper presented at the Society for the History of Technology conference, Washington, DC, October 17–21, 2007.

Pritchard, Sara B. "An Envirotechnical Disaster: Nature, Technology, and Politics at Fukushima." *Environmental History* 17, no. 2 (2012): 219–43.

Pritchard, Sara B. "Field Notes from the End of the World: Light, Darkness, Energy, and Endscape in Polar Night." *Journal of Energy History/Revue d'histoire de l'énergie* 2 (June 2019). https://www.energyhistory.eu/en/special-issue/epilogue-field-notes-end-world-light-darkness-energy-and-endscape-polar-night.

Pritchard, Sara B. "Joining Environmental History with Science and Technology Studies: Promises, Challenges, and Contributions." In *New Natures: Joining Environmental History with Science and Technology Studies*, edited by Dolly Jørgensen, Finn Arne Jørgensen, and Sara B. Pritchard, 1–17. Pittsburgh, PA: University of Pittsburgh Press, 2013.

Pritchard, Sara B. "Toward an Environmental History of Technology." In *The Oxford Handbook of Environmental History*, edited by Andrew C. Isenberg, 227–58. New York: Oxford University Press, 2014.

Pritchard, Sara B. "The Trouble with Darkness: NASA's Suomi Satellite Images of Earth at Night." *Environmental History* 22, no. 2 (2017): 312–30.

Pritchard, Sara B., Erin McLaughlin, and Michelle Shin. "Describing Artificial Light at Night: Keywords in Light Pollution Literature and Why They Matter" (unpublished manuscript).

Pritchard, Sara B., and Thomas Zeller. "The Nature of Industrialization." In *The Illusory Boundary: Environment and Technology in History*, edited by Stephen Cutcliffe and Martin Reuss, 69–100. Charlottesville: University of Virginia Press, 2010.

Proctor, Robert N., and Londa Schiebinger, eds. *Agnotology: The Making and Unmaking of Ignorance*. Stanford, CA: Stanford University Press, 2008.

Purdue, Peter. "Is There a Chinese View of Technology and Nature?" In *Illusory Boundary: Environment and Technology in History*, edited by Martin Reuss and Stephen H. Cutliffe, 101–19. Charlottesville: University of Virginia, 2010.

Pyne, Stephen J. *Between Two Fires: A Fire History of Contemporary America*. Tucson: University of Arizona Press, 2015.

Rader, Karen A. *Making Mice: Standardizing Animals for American Biomedical Research, 1900–1955*. Princeton, NJ: Princeton University Press, 2004.

Rajan, S. Ravi. *Modernizing Nature: Forestry and Imperial Eco-development, 1800–1950*. New York: Oxford University Press, 2006.

Rajaram, Vasudevan, Faisal Zia Siddiqui, and Mohd Emran Khan. *From Landfill Gas to Energy: Technologies and Challenges*. Boca Raton, FL: CRC, 2011.

Rand, Lisa Ruth. "Falling Cosmos: Nuclear Reentry and the Environmental History of Earth Orbit." *Environmental History* 24, no. 1 (2019): 78–103.

Rathje, William, and Cullen Murphy. *Rubbish! The Archaeology of Garbage*. New York: HarperCollins, 1992.

Raworth, Kate. "Must the Anthropocene be a Manthropocene?" *Guardian*, October 20, 2014. https://www.theguardian.com /commentisfree/2014/oct/20/anthropocene-working-group-science -gender-bias.

Redfield, Peter. *Life in Crisis: The Ethical Journey of Doctors Without Borders*. Berkeley: University of California Press, 2013.

Reid, Donald. *Paris Sewers and Sewermen: Realities and Representations*. Cambridge, MA: Harvard University Press, 1991.

Reno, Joshua O. *Waste Away: Working and Living with a North American Landfill*. Berkeley: University of California Press, 2015.

Reuss, Martin, and Stephen H. Cutcliffe, eds. *The Illusory Boundary: Environment and Technology in History*. Charlottesville: University of Virginia Press, 2010.

Reynolds, Terry S. *Stronger than a Hundred Men: A History of the Water Wheel*. Baltimore: Johns Hopkins University Press, 1983.

Richardson, Sarah S. *Sex Itself: The Search for Male and Female in the Human Genome*. Chicago: University of Chicago Press, 2013.

Roberts, Jody A., and Nancy Langston. "Toxic Bodies / Toxic Environments: An Interdisciplinary Forum." *Environmental History* 13, no. 4 (2008): 629–35.

Roeder, George H., Jr. "Coming to Our Senses." *Journal of American History* 81, no. 3 (1994): 1112–22.

Rome, Adam. *The Bulldozer in the Countryside: Suburban Sprawl and the Rise of American Environmentalism*. New York: Cambridge University Press, 2001.

Roque, Ricardo, and Kim A. Wagner, eds. *Engaging Colonial Knowledge: Reading European Archives in World History*. New York: Palgrave Macmillan, 2012.

Rosen, Christine Meisner. "Businessmen against Pollution in Late Nineteenth Century Chicago." *Business History Review* 69, no. 3 (1995): 351–97.

Rosen, Christine Meisner. "'Knowing' Industrial Pollution: Nuisance Law and the Power of Tradition in a Time of Rapid Economic Change." *Environmental History* 8, no. 4 (2003): 565–97.

Rosen, Christine Meisner. "The Role of Pollution Regulation and Litigation in the Development of the U.S. Meatpacking Industry, 1865–1880." *Enterprise and Society* 8, no. 2 (2007): 297–347.

Rosen, George. *A History of Public Health*. Revised ed. Baltimore: Johns Hopkins University Press, 2015.

Rozwadowski, Helen M. *Fathoming the Ocean: The Discovery and Exploration of the Deep Sea*. Cambridge, MA: Belknap Press of Harvard University Press, 2005.

Russell, Andrew L., and Lee Vinsel. "After Innovation, Turn to Maintenance." *Technology and Culture* 59, no. 1 (2018): 1–25.

Russell, Edmund. "Can Organisms Be Technology?" In *The Illusory Boundary: Environment and Technology in History*, edited by

Martin Reuss and Stephen H. Cutcliffe, 257–58. Charlottesville: University of Virginia Press, 2010.

Russell, Edmund. "Evolutionary History: Prospectus for a New Field." *Environmental History* 8, no. 2 (2003): 204–28.

Russell, Edmund. *Evolutionary History: Uniting History and Biology to Understand Life on Earth*. New York: Cambridge University Press, 2011.

Russell, Edmund. "Introduction: The Garden in the Machine; Toward an Evolutionary History of Technology." In *Industrializing Organisms: Introducing Evolutionary History*, edited by Susan R. Schrepfer and Philip Scranton, 1–16. New York: Routledge, 2004.

Russell, Edmund. "The Strange Career of DDT: Experts, Federal Capacity, and Environmentalism in World War II." *Technology and Culture* 40, no. 4 (1999): 770–96.

Russell, Edmund. *War and Nature: Fighting Humans and Insects with Chemicals from World War I to Silent Spring*. Cambridge: Cambridge University Press, 2001.

Russell, Edmund, James Allison, Thomas Finger, John K. Brown, Brian Balogh, and W. Bernard Carlson. "The Nature of Power: Synthesizing the History of Technology and Environmental History." *Technology and Culture* 42, no. 2 (2011): 246–59.

Sackman, Douglas C. "Nature's Workshop: The Work Environment and Workers' Bodies in California's Citrus Industry, 1900–1940." *Environmental History* 5, no. 1 (2000): 27–53.

Sackman, Douglas C. *Orange Empire: California and the Fruits of Eden*. Berkeley: University of California Press, 2005.

Sahlins, Peter. *Forest Rites: The War of the Demoiselles in Nineteenth-Century France*. Cambridge, MA: Harvard University Press, 1994.

Salehabadi, Djahane. "The Scramble for Digital Waste in Berlin." In *Re/Cycling Histories: Paths towards Sustainability*, edited by Ruth Oldenziel and Helmuth Trischler, 202–15. Oxford: Berghahn Books, 2016.

Sandlos, John, and Arn Keeling. "Living with Zombie Mines." *Seeing the*

*Woods* (Rachel Carson Center blog), March 6, 2013. https://seeing
thewoods.org/2013/03/06/living-with-zombie-mines/.

Sawyer, Richard C. *To Make a Spotless Orange: Biological Control in
California*. Ames: Iowa State University Press, 1997.

Schatzberg, Eric. "*Technik* Comes to America: Changing Meanings of
*Technology* before 1930." *Technology and Culture* 47, no. 3 (2006):
486–512.

Schatzberg, Eric. *Technology: Critical History of a Concept*. Chicago:
University of Chicago Press, 2018.

Schivelbusch, Wolfgang. *Disenchanted Night: The Industrialization
of Light in the Nineteenth Century*. Translated by Angela Davies.
Berkeley: University of California Press, 1988.

Schivelbusch, Wolfgang. *The Railway Journey: The Industrialization of
Time and Space in the 19th Century*. Berkeley: University of California Press, 1986.

Schmitt, Peter J. *Back to Nature: The Arcadian Myth in Urban America*.
New York: Oxford University Press, 1969.

Schneider, Daniel. *Hybrid Nature: Sewage Treatment and the Contra-
dictions of the Industrial Ecosystem*. Cambridge, MA: MIT Press,
2011.

Schneider, Mindi, and Philip McMichael. "Deepening, and Repairing, the
Metabolic Rift." *Journal of Peasant Studies* 37, no. 3 (2010): 461–84.

Schrepfer, Susan, and Philip Scranton, eds. *Industrializing Organisms:
Introducing Evolutionary History*. New York: Routledge, 2004.

Scott, James C. *Seeing like a State: How Certain Schemes to Improve the
Human Condition Have Failed*. New Haven, CT: Yale University
Press, 1998.

Sellers, Christopher C. *Hazards of the Job: From Industrial Disease to
Environmental Health Science*. Chapel Hill: University of North
Carolina Press, 1997.

Sellers, Christopher, and Joseph Melling, eds. *Dangerous Trade: Histories of Industrial Hazard across a Globalizing World*. Philadelphia:
Temple University Press, 2012.

Shallat, Todd. *Structures in the Stream: Water, Science, and the Rise of the U.S. Army Corps of Engineers*. Austin: University of Texas Press, 1994.

Shapiro, Judith. *Mao's War against Nature: Politics and the Environment in Revolutionary China*. New York: Cambridge University Press, 2001.

Shaw, Robert. "Night as Fragmenting Frontier: Understanding the Night That Remains in an Era of 24/7." *Geography Compass* 9, no. 12 (2015): 637–47.

Shulman, Peter A. *Coal and Empire: The Birth of Energy Security in Industrial America*. Baltimore: Johns Hopkins University Press, 2015.

Sinclair, Upton. *The Jungle*. New York: Doubleday, 1906. Reprinted, Oxford: Oxford University Press, 2010.

Smil, Vaclav. *Energy in World History*. Boulder, CO: Westview Press, 1994.

Smil, Vaclav, and Kazuhiko Kobayashi. *Japan's Dietary Transition and Its Impacts*. Cambridge, MA: MIT Press, 2012.

Smith, Andrew F. *Pure Ketchup: A History of America's National Condiment*. Columbia: University of South Carolina Press, 1996.

Smith, Mark M. *Sensing the Past: Seeing, Hearing, Smelling, Tasting, and Touching in History*. Berkeley: University of California Press, 2007.

Smith, Mark M. "Still Coming to 'Our' Senses: An Introduction." *Journal of American History* 95, no. 2 (2008): 378–80.

Smith, Neil. "There's No Such Thing as a Natural Disaster." Understanding Katrina Essay Forum. New York: Social Science Research Council, 2006. Available at *Items: Insights from the Social Sciences*. https://items.ssrc.org/understanding-katrina/theres-no-such-thing-as-a-natural-disaster/.

Smith, Neil. *Uneven Development: Nature, Capital, and the Production of Space*. New York: Blackwell, 1984.

Smith-Howard, Kendra. *Pure and Modern Milk: An Environmental History since 1900*. New York: Oxford University Press, 2014.

Soluri, John. "Accounting for Taste: Export Bananas, Mass Markets, and Panama Disease." *Environmental History* 7, no. 3 (2002): 386–410.

Soluri, John. *Banana Cultures: Agriculture, Consumption, and Environmental Change in Honduras and the United States.* Austin: University of Texas Press, 2005.

Solzman, David M. *The Chicago River: An Illustrated History and Guide to the River and Its Waterways.* Chicago: University of Chicago Press, 2006.

Spence, Mark David. *Dispossessing Wilderness: Indian Removal and the Making of the National Parks.* New York: Oxford University Press, 1999.

Stearns, Peter N. *The Industrial Revolution in World History.* 4th ed. New York: Routledge, 2012.

Steger, Brigitte. "'We were all in this together . . .': Challenges to and Practices of Cleanliness in Tsunami Evacuation Shelters in Yamada, Iwate Prefecture, 2011." *Asia-Pacific Journal* 10 (September 9, 2012). http://apjjf.org/2012/10/38/Brigitte-Steger/3833/article.html.

Steinberg, Theodore. *Acts of God: The Unnatural History of Natural Disaster in America.* 2nd ed. New York: Oxford University Press, 2006.

Steinberg, Theodore. *American Green: The Obsessive Quest for the Perfect Lawn.* New York: W. W. Norton, 2006.

Steinberg, Theodore. "An Ecological Perspective on the Origins of Industrialization," *Environmental Review* 10, no. 4 (1986): 261–76.

Steinberg, Theodore. *Nature Incorporated: Industrialization and the Waters of New England.* New York: Cambridge University Press, 1991.

Sterne, Jonathan, ed. *The Sound Studies Reader.* New York: Routledge, 2012.

Stine, Jeffrey K., and Joel A. Tarr. "At the Intersection of Histories: Technology and the Environment." *Technology and Culture* 39, no. 4 (1998): 601–40.

Stokes, Raymond, Roman Köster, and Stephen Sambrook. *The Business of Waste: Great Britain and Germany, 1945 to the Present.* New York: Cambridge University Press, 2013.

Stoler, Ann Laura. *Along the Archival Grain: Epistemic Anxieties and Colonial Common Sense.* Princeton, NJ: Princeton University Press, 2009.

Stoll, Steven. *The Fruits of Natural Advantage: Making the Industrialized Countryside.* Berkeley: University of California Press, 1998.

Stradling, David. *Smokestacks and Progressives: Environmentalists, Engineers, and Air Quality in America, 1881–1951.* Baltimore: Johns Hopkins University Press, 1999.

Stroud, Ellen. "Dead Bodies in Harlem: Environmental History and the Geography of Death." In *The Nature of Cities: Culture, Landscape and Urban Space,* edited by Andrew Isenberg, 62–76. Rochester, NY: University of Rochester Press, 2006.

Stroud, Ellen. "From Six Feet Under the Field: Dead Bodies in the Classroom." *Environmental History* 8, no. 4 (2003): 618–27.

Sutter, Paul S. *Driven Wild: How the Fight against Automobiles Launched the Modern Wilderness Movement.* Seattle: University of Washington Press, 2002.

Sutter, Paul S. "Nature's Agents or Agents of Empire? Entomological Workers and Environmental Change during the Construction of the Panama Canal." *Isis* 98, no. 4 (2007): 724–54.

Sutter, Paul S. "The World with Us: The State of American Environmental History." *Journal of American History* 100, no. 1 (2013): 94–119.

Suzik, Jeffrey Ryan. "'Building Better Men': Education, Training and Socialization of Working-Class Male Youth in the Civilian Conservation Corps, 1933–1942." PhD dissertation, Carnegie Mellon University, 2005.

Sze, Julie. *Noxious New York: The Racial Politics of Urban Health and Environmental Justice.* Cambridge, MA: MIT Press, 2007.

Tarr, Joel A. *The Search for the Ultimate Sink: Urban Pollution in Historical Perspective.* Akron, OH: University of Akron Press, 1996.

Tarr, Joel A., and Carl Zimring. "The Struggle for Smoke Control in St. Louis: Achievement and Emulation." In *Common Fields: The Environmental History of St. Louis*, edited by Andrew Hurley, 190–220. St. Louis: Missouri Historical Society, 1997.

Taylor, Dorceta E. *Toxic Communities: Environmental Racism, Industrial Pollution, and Residential Mobility*. New York: New York University Press, 2014.

Taylor, Joseph E., III. *Making Salmon: An Environmental History of the Northwest Fisheries Crisis*. Seattle: University of Washington Press, 1999.

Tenner, Edward. *Our Own Devices: The Past and Futures of Body Technologies*. New York: Knopf, 2003.

Tenner, Edward. *Why Things Bite Back: Technology and the Revenge of Unintended Consequences*. New York: Vintage, 1997.

Thompson, Emily. *The Soundscape of Modernity: Architectural Acoustics and the Culture of Listening in America, 1900–1933*. Cambridge, MA: MIT Press, 2002.

Thompson, Michael. *Rubbish Theory: The Creation and Destruction of Value*. New York: Oxford University Press, 1979.

Thoreau, Henry David. "Walking." *Atlantic Monthly* (June 1862): 657–74.

Thorsheim, Peter. *Inventing Pollution: Coal, Smoke, and Culture in Britain since 1800*. Athens: Ohio University Press, 2006.

Tobey, Ronald C. *Technology as Freedom: The New Deal and the Electrical Modernization of the American Home*. Berkeley: University of California Press, 1996.

Tompkins, Kyla Wazana. *Racial Indigestion: Eating Bodies in the Nineteenth Century*. New York: New York University Press, 2012.

Toynbee, Arnold. *Lectures on the Industrial Revolution in England: Public Addresses, Notes, and Other Fragments, Together with a Short Memoir by B. Jowett*. London: Rivington's, 1884.

Tsing, Anna Lowenhaupt. *The Mushroom at the End of the World: On the Possibility of Life in Capitalist Ruins*. Princeton, NJ: Princeton University Press, 2015.

Udall, Stewart L. *The Quiet Crisis.* New York: Holt, Rinehart & Winston, 1963.

Uekötter, Frank. *The Age of Smoke: Environmental Policy in Germany and the United States, 1880–1970.* Pittsburgh: University of Pittsburgh Press, 2009.

Valenčius, Conevery Bolton. *The Health of the Country: How American Settlers Understood Themselves and Their Land.* New York: Basic Books, 2002.

Valenčius, Conevery Bolton. *The Lost History of the New Madrid Earthquakes.* Chicago: University of Chicago Press, 2013.

Valentine, Ben. "Plastiglomerate, the Anthropocene's New Stone." *Hyperallergic* (November 25, 2015). http://hyperallergic.com /249396/plastiglomeratetheanthropocenesnewstone/.

Van Horssen, Jessica. *A Town Called Asbestos: Environmental Contamination, Health, and Resilience in a Resource Community.* Vancouver: University of British Columbia Press, 2016.

Van Horssen, Jessica, and Rhada-Prema McAllister. *Asbestos, PQ: A Graphic Novel.* London, Ontario: Western University, 2009. http:// www.megaprojects.uwo.ca.

Vileisis, Ann. "Are Tomatoes Natural?" In *The Illusory Boundary: Environment and Technology in History,* edited by Martin Reuss and Stephen H. Cutcliffe, 211–48. Charlottesville: University of Virginia Press, 2010.

Vogel, Sarah A. *Is It Safe? BPA and the Struggle to Define the Safety of Chemicals.* Berkeley: University of California Press, 2013.

Voyles, Traci Brynne. *Wastelanding: Legacies of Uranium Mining in Navajo Country.* Minneapolis: University of Minnesota Press, 2015.

Walker, Brett L. *Toxic Archipelago: A History of Industrial Disease in Japan.* Seattle: University of Washington Press, 2010.

Warren, Christian. *Brush with Death: A Social History of Lead Poisoning.* Baltimore: Johns Hopkins University Press, 2001.

Washington, Sylvia Hood. *Packing Them In: An Archaeology of Envi-*

*ronmental Racism in Chicago, 1865–1954*. Lanham, MD: Lexington Books, 2005.

Washington, Sylvia Hood, Paul C. Rosier, and Heather Goodall. *Echoes from the Poisoned Well: Global Memories of Environmental Injustice*. Lanham, MD: Lexington Books, 2006.

Waters, Colin N., et al. "The Anthropocene Is Functionally and Stratigraphically Distinct from the Holocene." *Science* 351, no. 6269 (January 8, 2016). https://science.sciencemag.org/content/351/6269/aad2622

Weintraubjan, Karen. "More Female Sea Turtles Born as Temperatures Rise." *New York Times* (January 10, 2018).

Wenz, Peter. "Just Garbage: The Problem of Environmental Racism." In *Faces of Environmental Racism: Confronting Issues of Global Justice*, edited by Laura Westra and Peter Wenz, 57–71. Lanham, MD: Rowman & Littlefield, 1995.

Werner, Dietrich, and William Edward Newton, eds. *Nitrogen Fixation in Agriculture, Forestry, Ecology, and the Environment*. New York: Springer, 2005.

Wheeler, Stephen M., and Timothy Beatley, eds. *The Sustainable Urban Development Reader*, 3rd edition. New York: Routledge, 2014.

White, Richard. "'Are You an Environmentalist or Do You Work for a Living?': Work and Nature." In *Uncommon Ground: Toward Reinventing Nature*, edited by William Cronon, 171–85. New York: W. W. Norton, 1995.

White, Richard. "From Wilderness to Hybrid Landscapes: The Cultural Turn in Environmental History," *Historian* 66 (2004): 557–64.

White, Richard. *The Organic Machine: The Remaking of the Columbia River*. New York: Hill & Wang, 1995.

Wilkinson, Richard G. "The English Industrial Revolution." In *The Ends of the Earth: Perspectives on Modern Environmental History*, edited by Donald Worster, 80–102. New York: Cambridge University Press, 1988.

Williams, Raymond. "Ideas of Nature." In *Problems in Materialism and Culture: Selected Essays*, 67–85. London: Verso, 1980.

Williams, Raymond. *Keywords: A Vocabulary of Culture and Society.* New York: Oxford University Press, 1976.

Williams, Raymond. *New Keywords: A Revised Vocabulary of Culture and Society.* Malden, MA: Blackwell, 2005.

Willis, Dan, William W. Braham, Katsuhiko Muramoto, and Daniel A. Barber, eds. *Energy Accounts: Architectural Representations of Energy, Climate, and the Future.* New York: Routledge, 2017.

Wilson, Daniel C. S. "Arnold Toynbee and the Industrial Revolution: The Science of History, Political Economy, and the Machine Past." *History and Memory* 26, no. 2 (2014): 133–61.

Winson, Anthony. *The Industrial Diet: The Degradation of Food and the Struggle for Healthy Eating.* Vancouver: University of British Columbia Press, 2013.

Wirth, John D. *Smelter Smoke in North America: The Politics of Transborder Pollution.* Lawrence: University Press of Kansas, 2000.

Wirth, John D. "The Trail Smelter Dispute: Canadians and Americans Confront Transboundary Pollution, 1927–1941." *Environmental History* 1, no. 2 (1996): 34–51.

Wormbs, Nina. "Eyes on the Ice: Satellite Remote Sensing and the Narratives of Visualized Data." In *When the Ice Breaks: Media and the Politics of Arctic Climate Change*, edited by Miyase Christensen, Annika E. Nilsson, and Nina Wormbs, 52–69. Hampshire, UK: Palgrave Macmillan, 2013.

Worster, Donald. *The Dust Bowl: The Southern Plains in the 1930s.* New York: Oxford University Press, 1979.

Worster, Donald. *Rivers of Empire: Water, Aridity, and the Growth of the American West.* New York: Oxford University Press, 1985.

Worster, Donald, ed. *The Ends of the Earth: Perspectives on Modern Environmental History.* New York: Cambridge University Press, 1988.

Wright, Miriam. *A Fishery for Modern Times: The State and the Industrialization of the Newfoundland Fisher, 1934–1968.* New York: Oxford University Press, 2001.

Wrigley, E. A. *Continuity, Chance, and Change: The Character of the Industrial Revolution in England.* Cambridge: Cambridge University Press, 1988.

Yan, Hong-Seng. *Reconstruction Designs of Lost Ancient Chinese Machinery.* Dordrecht, Netherlands: Springer, 2007.

Yochim, Michael J. *Yellowstone and the Snowmobile: Locking Horns over National Park Use.* Lawrence: University Press of Kansas, 2009.

Yong, Ed. *I Contain Multitudes: The Microbes within Us and a Grander View of Life.* New York: Ecco, 2016.

Young, James Harvey. *Pure Food: Securing the Pure Food and Drug Act of 1906.* Princeton, NJ: Princeton University Press, 1989.

Yusoff, Kathryn. *A Billion Black Anthropocenes or None.* Minneapolis: University of Minnesota Press, 2018.

Zeller, Thomas. *Driving Germany: The Landscape of the German Autobahn, 1930–1970.* New York: Berghahn Books, 2007.

Zimring, Carl A. *Aluminum Upcycled: Sustainable Design in Historical Perspective.* Baltimore: Johns Hopkins University Press, 2017.

Zimring, Carl A. *Cash for Your Trash: Scrap Recycling in America.* New Brunswick, NJ: Rutgers University Press, 2005.

Zimring, Carl A. *Clean and White: A History of Environmental Racism in the United States.* New York: New York University Press, 2015.

Zimring, Carl A. "The Complex Environmental Legacy of the Automobile Shredder." *Technology and Culture* 52, no. 3 (2011): 523–47.

Zimring, Carl A. "Dirty Work: How Hygiene and Xenophobia Marginalized the American Waste Trades, 1870–1930." *Environmental History* 9, no. 1 (2004): 90–112.

Zimring, Carl A., and Michael A. Bryson. "Infamous Past, Invisible Present: Searching for Bubbly Creek in the 21st Century." *IA: The*

*Journal of the Society for Industrial Archaeology* 39, nos. 1–2 (2013): 79–91.

Zohary, Daniel, and Maria Hopf. *Domestication of Plants in the Old World*, 3rd ed. Oxford: Oxford University Press, 2000.

Zola, Émile. *Germinal*. Paris: G. Charpentier, 1885.

# Index

drugs, release into the environment, 9, 133

Earth, photographed from space, 146–47
earthquakes, 103–5, 106, 109; anthropogenic, 50; as nuclear power plant disaster cause, 76, 99, 100, 101, 103, 109, 110–11, 113, 129
eating, as political act, 157
ecological changes, environmental generational amnesia and, 108–9, 112–13
ecosystems, relationship with human body, 8–9
ecosystem services, 121
electrification, 13
electronics, as waste, 76, 94, 95–97
embodied environmental history, 92, 126–28
empire, 4, 44–45. *See also* colonialization
endocrine disrupters, 1, 9, 37, 107, 128–29, 131, 132
Engels, Friedrich, 44, 65
environment: "clean" and "dirty" categorization, 45, 154; contexts, 3; embodied knowledge of, 92, 126–28; humans' effects on, 131–34
environmental determinism, 19–20
environmental generational amnesia, 108–9, 112–13
environmental historians, 1–2
environmental history, 2–3
environmental justice movement, 7, 11, 61–62, 92, 96, 130, 163

environmental protection, inequalities in, 60–62, 92
envirotechnical history, 129, 162–68
envirotechnical scholarship, 2, 4
envirotechnical systems, 7, 9–10, 15–16, 163
epidemics, 74, 90, 103–4, 105
ethnography, multispecies, 138
European Union, "extended producer responsibility" regulations, 97
evolution and evolutionary change, 46, 64, 107, 134, 149–50, 160

feminist studies, 64–65, 116–17
fertilizers, 28, 29–31, 40–41, 68, 153
fire/fires, 12, 20, 120
Flint water crisis, 1, 15, 127–28
floods, 8, 31, 63, 100
Food and Drug Administration (FDA), 37
food and food systems, 5–6, 17–42; canned foods, 26, 158–59; centralization, 27; Columbian exchange system, 21–22; definition, 18; environmental effects, 31–37; envirotechnical history, 163–64; genetic engineering and, 1, 38–41, 158, 160; organic and locavore, 41; packaging, 27–28; preindustrial, 18–22; prepackaged food, 35–36; processed foods, 17, 25–26, 39–40, 41, 158–59, 163; regulation, 36–37; storage and preservation methods, 24, 25–26; sustainable, 41–42; taste of food, 156–60; technological innovations of, 20–31; tools and machinery for, 34–36. *See also*